AF831414

W
I

Agric

DIGITAL MEDIA AND WIRELESS COMMUNICATIONS IN DEVELOPING NATIONS

Agriculture, Education, and the Economic Sector

Edited by

Megh R. Goyal

Emmanuel Eilu

AAP | APPLE ACADEMIC PRESS

Apple Academic Press Inc.
3333 Mistwell Crescent
Oakville, ON L6L 0A2, Canada

Apple Academic Press Inc.
1265 Goldenrod Circle NE
Palm Bay, Florida 32905, USA

© 2020 by Apple Academic Press, Inc.

Exclusive worldwide distribution by CRC Press, a member of Taylor & Francis Group

No claim to original U.S. Government works

International Standard Book Number-13: 978-1-77188-785-4 (Hardcover)
International Standard Book Number-13: 978-0-42902-931-8 (eBook)

All rights reserved. No part of this work may be reprinted or reproduced or utilized in any form or by any electric, mechanical or other means, now known or hereafter invented, including photocopying and recording, or in any information storage or retrieval system, without permission in writing from the publisher or its distributor, except in the case of brief excerpts or quotations for use in reviews or critical articles.

This book contains information obtained from authentic and highly regarded sources. Reprinted material is quoted with permission and sources are indicated. Copyright for individual articles remains with the authors as indicated. A wide variety of references are listed. Reasonable efforts have been made to publish reliable data and information, but the authors, editors, and the publisher cannot assume responsibility for the validity of all materials or the consequences of their use. The authors, editors, and the publisher have attempted to trace the copyright holders of all material reproduced in this publication and apologize to copyright holders if permission to publish in this form has not been obtained. If any copyright material has not been acknowledged, please write and let us know so we may rectify in any future reprint.

Trademark Notice: Registered trademark of products or corporate names are used only for explanation and identification without intent to infringe.

Library and Archives Canada Cataloguing in Publication

Title: Digital media and wireless communications in developing nations : agriculture, education, and the economic sector / edited by Megh R. Goyal, Emmanuel Eilu.

Names: Goyal, Megh Raj, editor. | Eilu, Emmanuel, editor.

Description: Includes bibliographical references and index.

Identifiers: Canadiana (print) 20190141662 | Canadiana (ebook) 20190141700 | ISBN 9781771887854 (hardcover) | ISBN 9780429029318 (ebook)

Subjects: LCSH: Digital media—Economic aspects—Developing countries. | LCSH: Wireless communication systems—Economic aspects—Developing countries. | LCSH: Developing countries—Economic conditions

Classification: LCC HM851 .D54 2019 | DDC 302.23/1091724—dc23

Library of Congress Cataloging-in-Publication Data

Names: Goyal, Megh Raj, editor. | Eilu, Emmanuel, editor.

Title: Digital media and wireless communications in developing nations : agriculture, education, and the economic sector / edited by Megh R. Goyal, Emmanuel Eilu.

Description: Oakville, ON, Canada ; Palm Bay, Florida : Apple Academic Press, [2020] | Includes bibliographical references and index. | Summary: "Digital Media and Wireless Communications in Developing Nations: Agriculture, Education, and the Economic Sector explores how digital media and wireless communications, especially mobile phones and social media platforms, offer concrete opportunities for developing countries to transform different sectors of their economies. The volume focuses on the agricultural, economic, and education sectors. The chapter authors, mostly from Africa and India, provide a wealth of information on recent innovations, the opportunities they provide, challenges faced, and the direction of future research in digital media and wireless communication to leverage transformation in developing countries. The volume presents a diverse selection research on such topics as In agriculture: mobile app that help to combat pests and diseases in tomatoes using mobile technology to prevent illegal timber logging In education: E-learning in adult education for prison inmates using social networks in health care education In economics: artificial intelligence to foster sustainable development, reforming public finance management via an integrated financial management system The volume provides important research on digital media and wireless communications within the context of developing countries that will be very useful for professionals from academia, government agencies, NGOs, technologists, entrepreneurs and investors, and others"-- Provided by publisher.

Identifiers: LCCN 2019026790 (print) | LCCN 2019026791 (ebook) | ISBN 9781771887854 (hardcover) | ISBN 9780429029318 (ebook)

Subjects: LCSH: Wireless communication systems--Developing countries. | Mobile communication system--Developing countries. | Agricultural innovations--Developing countries. | Mobile communication systems in education--Developing countries. | Mobile communication systems-Economic aspects--Developing countries.

Classification: LCC TK5103.2 .D535 2020 (print) | LCC TK5103.2 (ebook) | DDC 384.309172/4--dc23

LC record available at https://lccn.loc.gov/2019026790

LC ebook record available at https://lccn.loc.gov/2019026791

Apple Academic Press also publishes its books in a variety of electronic formats. Some content that appears in print may not be available in electronic format. For information about Apple Academic Press products, visit our website at **www.appleacademicpress.com** and the CRC Press website at **www.crcpress.com**

About the Senior Editor-in-Chief

Megh R. Goyal, PhD, PE

Retired Professor in Agricultural and Biomedical Engineering, University of Puerto Rico, Mayaguez Campus; Senior Acquisitions Editor, Biomedical Engineering and Agricultural Science, Apple Academic Press, Inc.

Megh R. Goyal, PhD, PE, is a Retired Professor in Agricultural and Biomedical Engineering from the General Engineering Department in the College of Engineering at the University of Puerto Rico–Mayaguez Campus; and Senior Acquisitions Editor and Senior Technical Editor-in-Chief in Agriculture and Biomedical Engineering for Apple Academic Press, Inc. He has worked as a Soil Conservation Inspector and as a Research Assistant at Haryana Agricultural University and Ohio State University.

During his professional career of 49 years, Dr. Goyal has received many prestigious awards and honors. He was the first agricultural engineer to receive the professional license in Agricultural Engineering in 1986 from the College of Engineers and Surveyors of Puerto Rico. In 2005, he was proclaimed as "Father of Irrigation Engineering in Puerto Rico for the Twentieth Century" by the American Society of Agricultural and Biological Engineers (ASABE), Puerto Rico Section, for his pioneering work on micro irrigation, evapotranspiration, agroclimatology, and soil and water engineering. The Water Technology Centre of Tamil Nadu Agricultural University in Coimbatore, India, recognized Dr. Goyal as one of the experts "who rendered meritorious service for the development of micro irrigation sector in India" by bestowing the Award of Outstanding Contribution in Micro Irrigation. This award was presented to Dr. Goyal during the inaugural session of the National Congress on "New Challenges and Advances in Sustainable Micro Irrigation" on March 1, 2017, held at Tamil Nadu Agricultural University. Dr. Goyal received the Netafim Award for Advancements in Microirrigation: 2018 from the American

Society of Agricultural Engineers at the ASABE International Meeting in August 2018.

A prolific author and editor, he has written more than 200 journal articles and textbooks and has edited over 59 books. He is the editor of three book series published by Apple Academic Press: Innovations in Agricultural & Biological Engineering, Innovations and Challenges in Micro Irrigation, and Research Advances in Sustainable Micro Irrigation. He is also instrumental in the development of the new book series Innovations in Plant Science for Better Health: From Soil to Fork.

Dr. Goyal received his BSc degree in engineering from Punjab Agricultural University, Ludhiana, India; his MSc and PhD degrees from Ohio State University, Columbus; and his Master of Divinity degree from Puerto Rico Evangelical Seminary, Hato Rey, Puerto Rico, USA.

About the Editor

Emmanuel Eilu, PhD

Emmanuel Eilu, PhD, is currently a Lecturer at Uganda Christian University in Mukono, Uganda. He holds a PhD in Information Technology from Makerere University in Uganda as well as an MS degree in Information Technology, specializing in E-government. Since 2007, Dr. Eilu has been teaching both postgraduate and undergraduate courses at the School of Computing and IT, Makerere University, Kampala, Uganda. Dr. Eilu has also worked as a Data Analyst Consultant for international NGOs. His areas of research and interest are in human computer interaction (usability, user experience, motivation, and acceptance), mobile user experience, and ICT for development (adoption of E-government systems; E-governance; E-voting; biometric voter registration systems; electronic civil registers, E-health and M-agriculture). His major teaching areas include usability and user experience, E-government, E-governance, Systems Analysis and Design, information systems security, information systems audit, data communication networks, and research methods. He has published numerous research papers.

Contents

About the Senior Editor-in-Chief by Megh R. Goyal v

About the Editor vii

Contributors xi

Abbreviations xiii

Preface xvii

PART I: Potential Use of Mobile Phones and Social Media in Agriculture... 1

1. **Potential of Mobile Phones as a Disruptive and Transformative Tool** **3**
   Hector John Mongi and Alfred Said Sife

2. **Transformative Power of Mobile Phones in the Indian Agro-Supply Chain** **25**
   George Joseph and K. A. Zakkariya

3. **Potential of Mobile Phones and Social Media Platforms in the Agricultural Scenario** **41**
   Kelvin Joseph Bwalya

4. **Role of Social Media in the Socioeconomic Development of Africa** **61**
   Kehbuma Langmia

5. **Mobile App-Based Expert System in Uganda: Combating Pests and Diseases in Tomatoes** **75**
   Emmanuel Eilu

6. **Role of Mobile Phone Technology in Combating Illegal Timber Logging in Africa: A Review** **91**
   Emmanuel Eilu

PART II: Selected Applications of Mobile Phones and Social Media in Education **103**

7. **Operational Use of Mobile Learning: Understanding and Retention of Curriculum Content** **105**
   Jeremy Sunker, Oyetade Kayode Emmanuel, and Kehdinga George Fomunyam

8. WhatsApp-Based E-Learning in Adult Education for Lesotho's Inmates: Methods and Evaluation 119
Selloane Pitikoe

9. Role of Teacher–Developer Relationship in Mobile Apps Learning..... 137
Soraya Sedkaoui

10. Integrating Social Media in Lesotho's Distance Learning: Gaps and Opportunities 157
Selloane Pitikoe

11. Role of Social Networks in Healthcare for Developing Countries...... 177
Jana Shafi, Amtul Waheed, and Sugapriya Dhanasekaran

PART III: Recent Strategies for Mobile Phones and Social Media in the Economic Sector 207

12. Potential of Streaming Transmission to Mobile Phone as Reinforcement of Community Radios 209
Celestino V. Joanguete

13. Luxury Brands on Social Media Platforms 225
Amit Kumar

14. Role of Artificial Intelligence in Sustainable Development................. 247
Mounia Khelfaoui

15. Mobile Technology Versus Era of Eco-Design 263
Mounia Khelfaoui and Soraya Sedkaoui

16. Reforming Public Finance Management in Africa Using Integrated Financial Management System 279
Emmanuel Eilu

Index 297

Contributors

Kelvin Joseph Bwalya, PhD
Associate Professor, Department of Information and Knowledge Management, School of Consumer Intelligence and Information Systems, College of Business and Economics, University of Johannesburg, Kingsway Campus, PO Box 524, Auckland Park, South Africa.
E-mail: kbwalya@uj.ac.za

Sugapriya Dhanasekaran, PhD
Assistant Professor, Department of Medical Laboratory Sciences (Pathology), College of Applied Medical Sciences, Prince Sattam Bin Abdulaziz University, Wadi-Al Dawaser Campus, Riyadh Province, Kingdom of Saudi Arabia.
E-mail: sughaphd@yahoo.com

Emmanuel Eilu, PhD
Lecturer (Information Technology), Faculty of Science and Technology, Uganda Christian University, Bishop Road P.O. Box 4, Mukono, Uganda.
E-mail: eiluemma@yahoo.co.uk

Oyetade Kayode Emmanuel, PhD Student
Lecturer, Department of Information Technology, Durban University of Technology, 41/43 M L Sultan Rd, Greyville, P.O. Box 1334, Durban 4001, South Africa.
E-mail; kayoyetade@gmail.com

Kehdinga George Fomunyam, PhD
Research Fellow, Department of Information Technology, Durban University of Technology, 41/43 M L Sultan Rd, Greyville, P.O. Box 1334, Durban 4001, South Africa.
E-mail; georgek@dut.ac.za

Megh R. Goyal, PhD, PE
Retired Faculty in Agricultural and Biomedical Engineering from the College of Engineering at University of Puerto Rico – Mayaguez Campus; Senior Acquisitions Editor and Senior Technical Editor-in-Chief in Agricultural and Biomedical Engineering for Apple Academic Press Inc.; PO Box 86, Rincon – PR – 006770086, USA. E-mail: goyalmegh@gmail.com

Celestino V. Joanguete, PhD
Researcher, School of Communication and Arts (ECA), Eduardo Mondlane University, Av. Romão Fernandes Farinha, nº 456 – 520 Cidade de Maputo, Maputo, Mozambique.
E-mail: celestino.joanguete@gmail.com

George Joseph, PhD
Assistant Professor, DDU Kaushal Kendra (DDUKK), Cochin University of Science and Technology (CUSAT), Ernakulam 682022, Cochin, Kerala, India. E-mail: cgeorgejoseph@gmail.com

Mounia Khelfaoui, PhD
Senior Lecturer (HDR), Department of Economics, Khemis Miliana University – Algeria, Route Thineyet El Had – Khemis Miliana 44225, Algeria; Researcher at Organizational Development and Innovation Laboratory – University of Khemis Miliana, Algeria.
E-mail: mouniakhelfaoui@yahoo.fr

Amit Kumar, PhD
Assistant Professor, Institute of Business Management and Research, IPS Academy, Affiliated Devi Ahilya University, A.B Road, Rajendra Nagar, Bijalpur 452012, Indore, Madhya Pradesh, India. E-mail: amitkumariccs@gmail.com

Kehbuma Langmia, PhD
Professor and Chair, Full-bright Scholar, Department of Strategic, Legal and Management Communication (SLMC), Cathy Hughes School of Communications, Howard University, 525 Bryant Street NW, Washington, DC 20059, USA. E-mail: klangmia@howard.edu

Hector John Mongi, PhD
Lecturer, University of Dodoma, College of Informatics and Virtual Education, P. O. Box 490, Dodoma, Tanzania. E-mail: hjmongi@yahoo.com

Selloane Pitikoe, MEd, PhD
University of Eswatini, Faculty of Education, Department of Adult Education, Room C.3.6, Lozitha road. Private Bag, 4, Kwaluseni M201, Eswatini.
E-mail: mofumahali@gmail.com

Soraya Sedkaoui, PhD
Senior Lecturer (HDR), Department of Economics, Khemis Miliana University – Algeria, Route Thineyet El Had –Khemis Miliana 44225, Algeria; Researcher at TRIS Laboratory – University of Montpellier; Data Analyst, SRY – Consulting, Montpellier, France.
E-mail: soraya.sedkaoui@gmail.com; sryconsulting1@gmail.com

Jana Shafi, MTech
Lecturer, Department of Computer Science, College of Arts and Science, Prince Sattam Bin Abdulaziz University, Wadi-Al Dawaser Campus, Riyadh Province, Kingdom of Saudi Arabia.
E-mail: janashafi09@gmail.com

Alfred Said Sife, PhD
Deputy Vice Chancellor (Academic), Moshi Co-Operative University (MoCU), Sokoine Road, P.O. Box 474, Moshi, Tanzania – East Africa; Former Associate Professor, Sokoine University of Agriculture, P. O. Box 3000 Chuo Kikuu, Morogoro, Tanzania. E-mail: sifeas@gmail.com

Jeremy Sunker, PhD Student
Department of Information Technology, Durban University of Technology, 41/43 M L Sultan Rd, Greyville, P.O. Box 1334, Durban 4001, South Africa. E-mail; jeremysunker@gmail.com

Amtul Waheed, MTech
Lecturer, Department of Computer Science, College of Arts and Science, Prince Sattam Bin Abdulaziz University, Wadi-Al Dawaser Campus, Riyadh Province, Kingdom of Saudi Arabia. E-mail: w_amtul@yahoo.com

K. A. Zakkariya, PhD
Professor and Director, DDU Kaushal Kendra (DDUKK), School of Management Studies, Cochin University of Science and Technology (CUSAT), Ernakulam 682022, Cochin, Kerala, India.
E-Mail: zakkariya@gmail.com

Abbreviations

ACC	advanced audio coding
AHRQ	Agency for Healthcare Research and Quality
AI	artificial intelligence
AIDS	acquired immune deficiency syndrome
AO	agricultural officer
BoP	base (or bottom) of the pyramid
BSNL	Bharat Sanchar Nigam Limited
CoA	chart of accounts
CODDE	Conception Development Durable Environment
CR	community radios
CRM	customer relationship management
CSA	climate smart agriculture
CSOs	Civil Society Organizations
DALY	disability-adjusted life-year
DFID	Department for International Development
DJ	disc jockey
DTEP	Distance Teacher Education Program
EAC	East African Community
ECLAC	Economic Commission for Latin America and the Caribbean
ED	eco-design
EMix	Electronic Medical Information Exchange
ER	eco-responsible
ERMT	eco-responsible mobile technology
ES	expert systems
FAO	Food and Agricultural Organization
FDA	US Food and Drug Administration
FDI	Foreign Direct Investment
FFO	Formalized Farmers Organization
FIG	farmers interest groups
FLAC	Free Lossless Audio Codec
FM	frequency modulation

FPO	Farmer Producer Organizations
GDP	gross domestic project
GPS	global positioning system
GSMA	Global System for Mobile Communications Association
GUI	graphical user interface
HbA1C	glycated hemoglobin
HCA	Hospital Corporation of America
HCAHPS	Hospital Consumer Assessment of Healthcare Providers and Systems
HDR	Human Development Reports
HE	higher education
HHS	US Health and Human Services
HIE	horizontal information exchanges
HIV	human immunodeficiency virus
HTML	hypertext mark-up language
ICASA	Independent Communications Authority of South Africa
ICT	Information and Communications Technology
IDA	International Development Association
IEMS	Institute of Extra-Mural Studies
IFFCO	Indian Farmers Fertilizer Cooperative Limited
IFMS	Integrated Financial Management Systems
ILDF	Integrative Learning Design Framework
IMF	International Monetary Fund
IPTV	Internet Protocol Television
ISFM	Integrated Soil Fertility Management
ISO	International Organization for Standardization
IT	information technology
ITU	International Telecommunication Union
IUCEA	Inter-University Council for East Africa
IVRS	interactive voice response
KVK	Krishi Vigyan Kendra
LCA	life-cycle assessment
LCE	Lesotho College of Education
LCS	Lesotho Correctional Services
LDTC	Lesotho Distance Teaching Centre
LVB	Lake Victoria Basin
LVBC	Lake Victoria Basin Commission
MDA	ministries, departments and agencies

MDGs	millennium development goals
M-health	mobile health
MMS	multimedia message service
MoET	Ministry of Education and Training
MSM	men-who-have-sex-with-men
MSSRF	M. S. Swaminathan Research Foundation
MT	mobile technology
mTh	Methodological Trojan Horse
NAIP	National Agricultural Innovation Project
NEPAD	New Partnership for Africa's Development
NFE	non-formal education
NGOs	non-governmental organizations
NMAET	National Mission on Agricultural Extension and Technology
NTTC	National Teachers Training College
NUL	National University of Lesotho
OS	operating system
PDA	personal digital assistants
PFM	public finance management
PSR	patient satisfaction ratings
PWC	Pricewaterhouse Coopers
R&D	research and development
RISDP	Regional Indicative Strategic Development Plan
RoT	rule of thumb
SADAG	South African Depression and Anxiety Group
SADC	Southern African Development Community
SD	sustainable development
SDGs	sustainable development goals
SHG	self-help group
Sida	Swedish International Development Agency
SITA	State Information Technology Agency
SMS	short message service
SNS	social networking sites
SRELT	simple relevance evaluation logic template
SSA	Sub-Saharan Africa
SSCM	specific crop management
TDT	transactional distance theory
TESOL	teaching English as second official language

UCC	Uganda Communication Commission
UDOM	University of Dodoma
UK	United Kingdom
UNCCD	United Nations Convention to Combat Desertification
UNDP	United Nations Development Programme
UNESC	United Nations Economic and Social Council
UNESCO	United Nations Education, Scientific and Cultural Organization
UP	University of Pretoria
URA	Uganda Revenue Authority
URL	Uniform Resource Locator
URT	United Republic of Tanzania
VAS	value-added service
VicRes	Lake Victoria Research Initiative
VIE	Vertical Information Exchanges
VO	Veterinary Officer
WHO	World Health Organization
WOUGNET	Women of Uganda Network
WRM	Water Resources Management
WSIS	World Summit for Information Society

Preface

Digital Media and Wireless Communications in Developing Nations: Agriculture, Education, and the Economic Sector explores how digital media and wireless communication technologies (like mobile phones and social media platforms) offer concrete opportunities for developing countries to transform different sectors of their economies, namely, the agricultural, education, and economic sectors.

This book volume, in short, explores how recent use of mobile phones and social media technologies has transformed the agricultural, education, and the economic sectors in developing countries. The volume provides a wealth of information on recent innovations, shortcomings, and future challenges of using mobile phones and social media technologies to leverage transformation in different sectors in developing countries. The book is divided into the following sections:

- an exploration of the use of mobile phones and social media in agriculture;
- recent innovations on the use of mobile phones and social media in education and learning;
- current applicability of mobile phones and social media in economic sector.

The information provided here will be valuable to government agricultural professionals, academia, governments, civil society organizations, technologists, researchers, farmers, and students all over the world. Key features of this book are as follows:

- It provides a wealth of information on recent innovations, shortcomings, and future challenges of using mobile phones and social media technologies to leverage transformation in the different sectors.
- It provides specific examples of the use of mobile phones and social media in different sectors.

The contributions by the chapter authors to this book volume have been most invaluable in the compilation. Their names are mentioned in each chapter and in the list of contributors. This book would not have been written without the valuable cooperation of these investigators, many of whom are renowned scientists who have worked in the field of information and communication technology throughout their professional careers.

The goal of this book is to guide the world science community on application of mobile phones, tablets, and social media in various sectors in developing countries.

We thank Ashish Kumar, Publisher and President at Apple Academic Press, Inc., for making every effort to publish the book when mobile technology has become an essential part of our life worldwide. Special thanks are due to the AAP Production Staff for their work.

We express our deep admiration to our families for understanding and collaboration during the preparation of this book.

As educators, we give a piece of advice to one and all in the world: "*Permit that our Almighty God, our Creator, allows us to inherit new technologies for a better life at our planet. I invite our community in agricultural engineering to contribute book chapters to our future books by getting married to our profession....*" We are in total love with our profession by length, width, height, and depth. Are you?

—Megh R. Goyal, PhD, PE
Emmanuel Eilu, PhD

PART I

Potential Use of Mobile Phones and Social Media in Agriculture

CHAPTER 1

Potential of Mobile Phones as a Disruptive and Transformative Tool

HECTOR JOHN MONGI and ALFRED SAID SIFE

ABSTRACT

Mobile technology has become ubiquitous in developing countries, where the rate of user penetration is amazing. The context of this booming tool has led to the emergence of two buzz words: disruptive and transformative technology. This chapter presents and discusses the results where the disruptive and transformative power of mobile phone was studied in the Lake Victoria Basin (LVB) of Tanzania. The study applied the context lens of selected small-scale Formalized Farmers Organizations at the bottom of the pyramid of the Tanzanian management structure of LVB. A theoretical approach that linked tool, activity, and goal was used to guide the study. Findings are used to discuss the mobile phone as a disruptive technology and as a tool for bringing faster transformation in the social, economic, and ecological dimensions of sustainable small-scale agriculture in the study area.

1.1 INTRODUCTION

Penetration and usage of mobile phones in the developing countries is amazing. The comparisons made on mobile phone user penetration between developed and developing economies in the world showed that the former had a significant leap into the better position.[34] For example, according to the 2015 estimates by the International Telecommunications Union (ITU), 96 own mobile phones for every 100 global citizens.[18] Much of this data results from rapid penetration in China and India, homes to more than a third of the global population.[43] Apart from the general trends, user penetration rate of smart phones has shown a drastic leap in some developing countries. For example, in India, penetration increased from 12.8% in 2013 to 51% in 2017.[11,34] The usage in these regions is becoming ubiquitous with ownership per capita that bypasses other indicators of possession of assets including per capita income. It is now evident that there is massive penetration of mobile phones in the developing world. The size of economies is, therefore, not the only determinant of mobile phone penetration.

Mobile phones contribute to overcoming some of the barriers to technology adoption especially in the developing world, which includes technological complexity, inadequate awareness, high investment cost, age, lack of network infrastructure, and perceived low trust.[17,22,25,49] Consequently, mobile phones are among the rapidly penetrating digital technologies urgently needed to accelerate the speed of addressing societal challenges. Two buzz words have emerged out of debates on the ecosystem of mobile phones in the developing world: disruptive and transformative technology. The two words are worth exploring further in the developing countries as part of the studies on the ubiquitous mobile phones.

This chapter presents and discusses the results, where the disruptive and transformative powers of the mobile phone were studied in the Lake Victoria Basin (LVB) of Tanzania. Two key research questions are answered: (1) How mobile phones are supporting the engagement of small-scale irrigation farmers within Formalized Farmers Organizations (FFOs) in the LVB of Tanzania with other actors in water resources management? and (2) To what extent does the use of mobile phone services promotes the social, economic, and environmental goals of the FFO in the study area? Furthermore, the study applied the context lens of selected small-scale FFO at the bottom of the pyramid (BoP) of the Tanzanian water resource management structure. The chapter is narrative in nature with the substantial use of processed primary data collected from the field. The principle of on-demand overview and alertness as well as on-going debates about these powers are presented and used to shape the discussions.

1.1.2 THE PRINCIPLE OF ON-DEMAND OVERVIEW AND ALERTNESS

The principle of on-demand overview and alertness refers to situation-driven collection of information, where human is assisted with electronic devices. The principle embodies the concept of LifeLogging. Stefan Selke[39] explains by mimicking the army, where the LifeLog concept originated. He introduces a principle called "Total overview and alertness." The soldier in the army becomes fully equipped in the helmet with different information collecting devices (for image, in-house and external sounds detection and capture); hence, the word "total." In this fashion, soldier becomes a human sensor. While the sensors and associated devices in a soldier's helmet are collecting data on a continuous basis, a mobile handset replaces them all, and for the convenience, it is used on-demand. With mobile phones, data are not collected all the time but on an occasional or on-demand basis. This is also applicable in the resource-constrained areas, where a mobile phone or any other data collecting device cannot be active all the time. In the rural areas of developing countries, there are still challenges related to network and electric power infrastructure, limited financial resources to cover costs of airtime as well as limited device capacity to store and transmit large amount data.[10,37]

1.2 DEBATES ON TRANSFORMATIVE AND DISRUPTIVE POWER OF MOBILE TECHNOLOGY

Debates on how mobile technology—especially the phones—contribute to development (M4D) has largely stemmed from its disruptive and transformative powers.[13,14,35,48,50] On one hand, debates about disruptive nature of mobile phones have focused on their ability to dismantle the hierarchical and bureaucratic information flow. That is the ability to create a common platform for sharing information. On the other hand, arguments on transformative nature of mobile phones have focused on their ability to bring about changes in social, economic, and environmental aspects.

Traditionally, many debates have composed proposing and opposing sides. Likewise, the ongoing debates about ICT for Development (ICT4D) including the Mobile for Development (M4D) have not avoided the effects of traditional nature of arguments. However, despite having proponents and opponents, this chapter refers the two sides as optimists and pessimists, respectively. In other words, the debates compose of those who see the light at the end of the tunnel (optimists) and those who need to be envisioned to achieve the same (pessimists). Optimists believe that developing countries need to embrace digital-driven transformation to better social, economic, and environmental conditions. For example, the World

Economic Forum (WEF) advocates for digital transformation based on four reasons: (1) inefficiency and poor delivery of services, (2) corruption that stalk the economic development, (3) widening gap of digital skills capacity, and (4) a widening gap between traditional economy and the digital economy.[40]

Other examples of optimistic view about the disruptive and transformative power of mobile phones come from the agricultural sector. One of the many aspects of these powers is high penetration rate and massive adoption in rural areas, which in the context of developing countries are homes to the majority of agricultural dwellers.[31] Others are the ability to connect farmers to the markets, identify crop diseases, verifying the genuine agricultural inputs such as fertilizers, pesticides, and seeds as well as providing other extension services.[3,24,29,49]

Despite this positive side of the debate, some authors have indicated divergence from conclusions made about the disruptive and transformative power of mobile phones. While some have consistently cautioned about possible falsified credits about mobile phone progression in developing countries, others have criticized the methodology that has been followed to arrive at the conclusions. They argue that contexts of studies have been too broad to the extent that actual local situations were overshadowed. In other words, the local contexts were not fully reflected in the studies. For example, the study in Kerala, India[19] revealed that mobile phones were transforming the way fishermen were connected to the markets. With their phones, the fishermen were able to identify the potential customers, the volume of the commodity in demand, aggregate their merchandise, and eventually deliver to the destination. However, nearly 10 years later, the conclusion made was criticized by arguing that the methodology was too broad to merit the very specific case of Kerala.[41] There are many such debates in the literature, which contributes to shaping our understanding of ICT4D as well as M4D.[45]

This chapter also contributes to the debates in a case study, where the local context of activities, tools, and goals were considered in understanding the disruptive and transformative power of mobile phones.

1.3 ACTIVITY–TOOL–GOAL APPROACH

Both disruption and transformation involve the human, technology, and goal factors. An activity–tool–goal approach was used to guide the study.

Although some work has been done at the interface of activities and ICT tools, yet the proliferation of such tools and their relation to activities at the BoP has not been fully explored. The missing link within the three components was reported as among the failure reasons of ICT projects in Africa and the developing world, in general, to achieve rapid disruption and transformation.[8,15] Apart from other support services that can put ICT tools to use such as availability of electric power, training, and maintenance, their worthiness in terms of attaining the goals is paramount. Figure 1.1 shows the conceived relationship between the "activity," "tool," and "goal" in the context of small-scale farming communities.

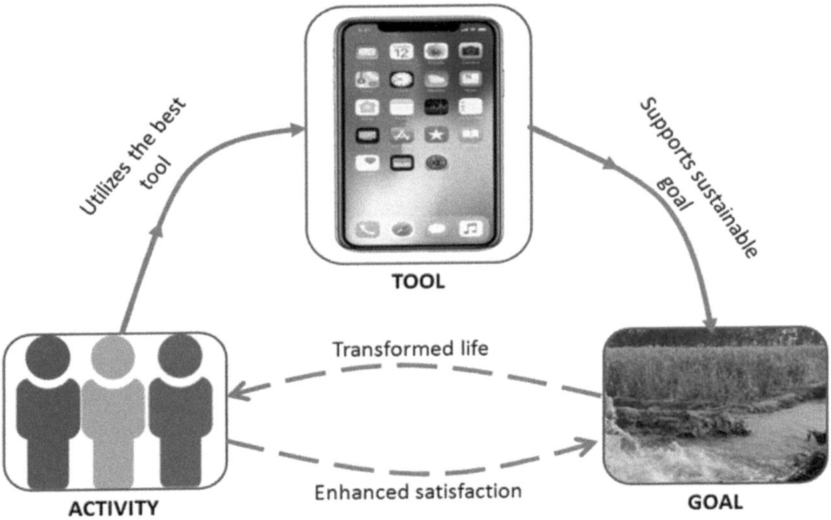

FIGURE 1.1 (See color insert.) The conceived nexus of FFO activities, the tool and goal.

A small-scale farming community finds a suitable communication tool to support them and achieve the goal in a sustainable manner. The relationship is further elaborated below where actual activities, tools, and goals are exemplified.

1.3.1 THE ACTIVITY

Human activities have been recorded as precursors of most sustainability concerns over the use of water resources for agriculture. This indicates a relationship between anthropogenic activities and sustainability; therefore, the need for their sustainable management.[38] Activities (A) conducted by FFO can be operationalized into one or more variables as shown in eq 1.1:

$$A = \{A_1, A_2 \ldots A_n\} \tag{1.1}$$

where n = number of elements making a set of activities, A_1 = planning, A_2 = policymaking, A_3 = decision-making, A_4 = law enforcement, A_5 = mobilization of financial resources, A_6 = infrastructure maintenance, A_7 = capacity building through education and training, and A_8 = other activities.

Each of these variables can represent a specific water resource management activity. For the purpose of this study, activities by members of small-scale irrigation organizations were derived from the local policy, strategies, and action plans. Furthermore, confirmation that they apply in the context of the study was checked during the first field visits. The following eight (n = 8) generic activities were therefore confirmed and included in the study.

1.3.2 THE TOOL

Researchers[1 4,27,30] contend that when a "tool" is designed in close collaboration with users when focused on their real needs, and when the cost issues and other technical barriers have been considered, then such a tool can be owned and be sustained. This study viewed the mobile solutions as having two interrelated and self-supporting parts: the social sub-systems and technology sub-systems. The social sub-systems part comprises the users, their activities in supporting the water resources, and their goals to ensure that sustainability of the resources is enhanced. The technology sub-systems part composes of mobile phone solutions for supporting various engaging activities in water resource management. The ICT tools (T) can also be operationalized into one or more variables as shown in eq 1.2:

$$T = \{T_1, T_2 \ldots T_n\} \tag{1.2}$$

where n = number of elements making a set of mobile phone tools (for the purpose of this study, four (n = 4) mobile phone service tools were perceived to be basic applications at BoP; T_1 = short message services (SMS), T_2 = voice messages, T_3 = images, and T_4 = video.

SMS and voice messages are possible for both the low-end (with limited features) and the high-end mobile phone (with many features). However, images and videos are tools common to the later.

1.3.3 THE GOAL

One of the goals of small-scale FFO is to attain the social, economic, and ecological dimensions of sustainable water resources. The Goal (G) can also be operationalized into one or more variables as shown in eq 1.3:

$$G = \{G_1, G_2 \ldots G_n\} \qquad (1.3)$$

Where n = number of elements contributing to an overall goal.

For the purpose of this study, sustainability of water resources was measured by indicators derived from both the Millennium Development Goals (MDGs) as well as by validation through the local experiences. The selected indicators as perceived by the small-scale FFO are shown in Table 1.1.

The validated indicators (Table 1.1) were ranked using 3-point Likert scale: weak, medium, strong. The indicators were also grouped into categories reflecting the three sustainability dimensions: social, economic, and ecological. Nearly equivalent indicators were identified from the MDGs as standard metrics. Most of the perceived indicators appeared to correspond with the standard ones, except for the differences in priorities perceived by respondents at this bottom level. Analysis revealed that although the identified indicators were not exactly as stated in the MDGs, close equivalent could be tracked (Table 1.2).

In summary, the following indicators that were used as proxies for measuring achievement of sustainability goal (G) for water resources in the study area were considered:

G_1 = Proportion of urban human population living in slums (the lower the better)

G_2 = Access to improved sanitation (the more the better)

G_3 = Amount of greenhouse gases production controlled

G_4 = Proportion of human population encroaching the water resources (the lower the better)

TABLE 1.1 Ranked Score for the Locally Perceived Goal Indicators.

Perceived goal indicator	Responses				Remarks
	Total scores	**Category**	**%**	**Cumulative %**	
Controlled livestock population	19	Economic	2.0	2	
Controlled illegal fishing within the resources	30	Ecological	3.1	5.1	
Regular training and advisory services	35	Economic	3.7	8.8	Weak
Reduced income poverty	34	Economic	3.5	12.3	
Regularly maintained infrastructures	61	Economic	6.3	18.6	
Controlled siltation of water reservoirs	82	Ecological	8.5	27.1	
Controlled human population	109	Social	11.2	38.3	
Controlled water pollution	142	Ecological	14.6	52.9	Medium
Controlled drought and floods	146	Ecological	15.1	68.0	
Mapped water resources and surroundings	149	Economic	15.4	83.2	Strong
Enforced rules and regulations	163	Social	16.8	100.0	
Total score	**970**		**100**		

TABLE 1.2 Fitting Perceived Local Indicators of Water Resources Sustainability into MDGs.

Priority local perceived indicator of water resource sustainability	Nearly equivalent MDG indicator related to water resource sustainability
Controlled drought and floods	Virtual elimination of ozone-depleting substances
Controlled human population	The proportion of urban population living in slums
Controlled water pollution	Access to improved sanitation
Enforced rules and regulations	All apply
Mapped water resources and surroundings	The proportion of urban population living in slums[a]

[a]In rural areas, it was related to encroachment of water sources.

1.4 POTENTIAL DISRUPTIVE AND TRANSFORMATIVE POWER OF MOBILE PHONE IN THE LVB

1.4.1 WATER RESOURCE MANAGEMENT IN THE LVB TANZANIA

The water resources management point of view, especially in the LVB, identifies three main management levels and four categories of water users. LVB is a shared water basin among five riparian states: Tanzania, Kenya, Uganda, Rwanda, and Burundi (Fig. 1.2).

LVB can be defined at regional management level, where the East Africa Community (EAC) discharges the regional water management functions. The regional economic block is an umbrella of organizations working at this level including the LVB Commission and others. Each of the five countries has management units for the water bodies falling in the state of jurisdictions. Singling Tanzania, where this case has been derived, LVB Tanzania constitutes one of the nine national level water bodies. LVB Tanzania is therefore managed at the local basin level with about five government entities known as regions: Mwanza, Kagera, Mara, Shinyanga, and Simiyu. Each of these regions has management functions discharged at Regional Secretariat for environmental affairs; District management functions discharged at District Councils and the Community levels (BoP).

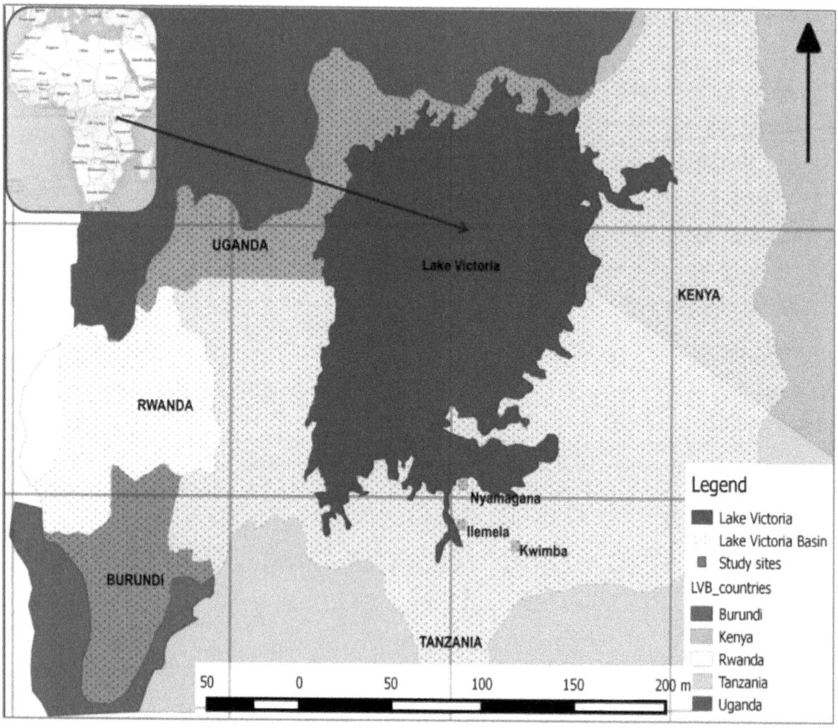

FIGURE 1.2 **(See color insert.)** Map of LVB showing the location of study sites in Tanzania.

Source: Authors' creation with Google Maps.

1.4.2 POTENTIAL DISRUPTIVE POWER

Small-scale FFO communicate among themselves as well as among occupants within the three scales of the administrative hierarchy. The scales have been referred to as micro, meso, and macro.[9,42,47] Horizontal information exchanges (HIE) and vertical information exchanges (VIE) within the three scales are graphically illustrated in Figure 1.3.

Micro or BoP is the foundational or community level where the potential impacts through the involvement of critical mass can be realized.[9,44] BoP can have different contextual descriptions: by location, by demography, by the administration, or by the participatory hierarchy. The World Bank, for example, describes BoP scale by income as that belonged to "persons who earn less than $2.50 per day."[7] The study area is considered as BoP of water resource users with multiple characteristics including those related to location, demography, and administrative hierarchy. Meso scale is an intermediate level with major tasks of organizing and coordinating.[9,47] Authors mention a meso scale as the tactical level which translates long-term plans into medium-term plans. Finally, a macro scale is mainly concerned with policy formulation, long-term planning, and development of legal and regulatory frameworks.[9,42]

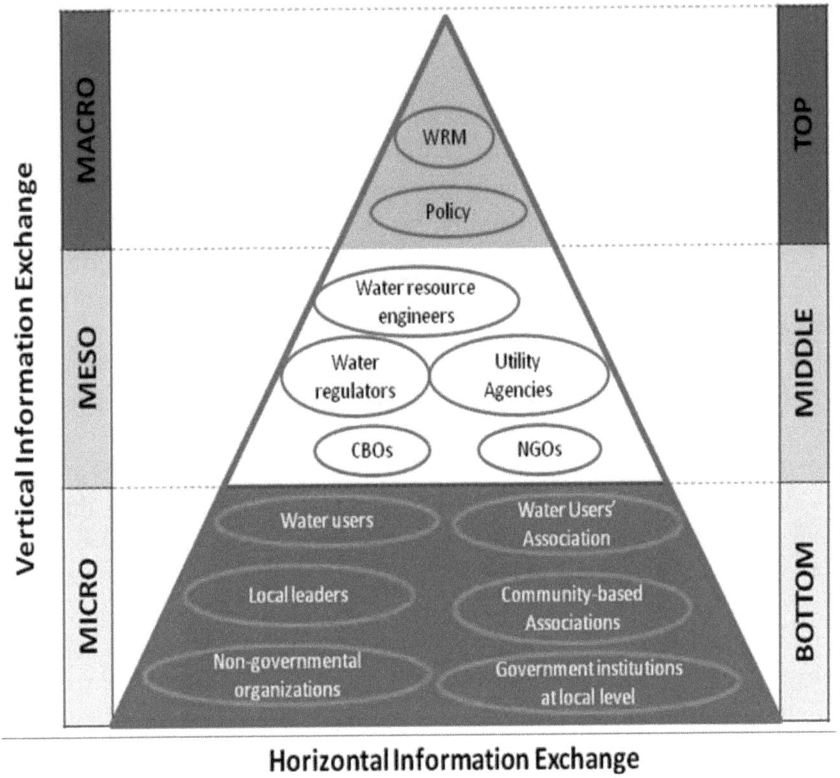

FIGURE 1.3 (See color insert.) The VIE and HIE for the three scales of WRM.

Effective VIE and HIE result to the engaged community. Understanding engagement at various scales is preceded by understanding the drivers behind it. The National Consumer Council (NCC) points out that the major driver of engagement is the values it can bring including helping to create better public services, promote social cohesion, and foster people-centered solutions.[33] The Council further argues that fully engaged members of the public have the potential to improve the quality of decision and policy solutions as well as enhancing democracy. Finally, it highlights the richness of data as a pre-requisite for valuable contribution to policy and decision-making by engaged members of the public. Such data, according to NCC, help to reveal public attitudes and values, offer opportunities to fully explore why people feel the way they do, and allow the time for development of ideas, options, and priorities with the public.

Community engagement allows for the inclusion of a wide range of people and information sources which have the potential to yield information, evidence, and views from people who view issues at different angles, who have varying backgrounds and shared interests. In order to maximize the values, public engagement at BoP must also be maximized, a situation which is yet to be attained in the water basin management of a developing country. It is therefore envisaged that such improvements disrupt the hierarchy of communication flow resulting into hypothetical low VIE (Fig. 1.4).

BoP is recognized for its potential for engagement of a wider horizon of the public in decision and policy-making in WRM-related issues. Chen et al.[2] argue that public engagement is fundamentally tied to the social capital, which describes as "social ties that bind communities and individuals together while producing a range of positive benefits that make vibrant communities healthier and resistant to change."[46] It is further argued that public engagement and social capital are related through the role of community-based activities in developing new or renewed ties within the same community (a "bringing together" process), assistance in the development of shared understanding and values, or through the aggregation and exchange of resources held within the community. Information is one of the critical resources characterizing the information society age.[2]

Despite the importance of the three scales in supporting engagement, VIE is limited by bureaucracy characterized by the structural hierarchy. Two authors, Harriden[12] and Lugo,[26] show that top-down approaches to transboundary solutions of water resource problems and politics associated with them have side-lined the stakeholders at BoP, thereby compounding the problem. The findings by Mongi and Meinhardt[32] from Southern Africa indicated that there was an improvement in stakeholders' involvement in design and development of information systems (disruptive forces) including those which uses mobile phones for WRM (Fig. 1.5).

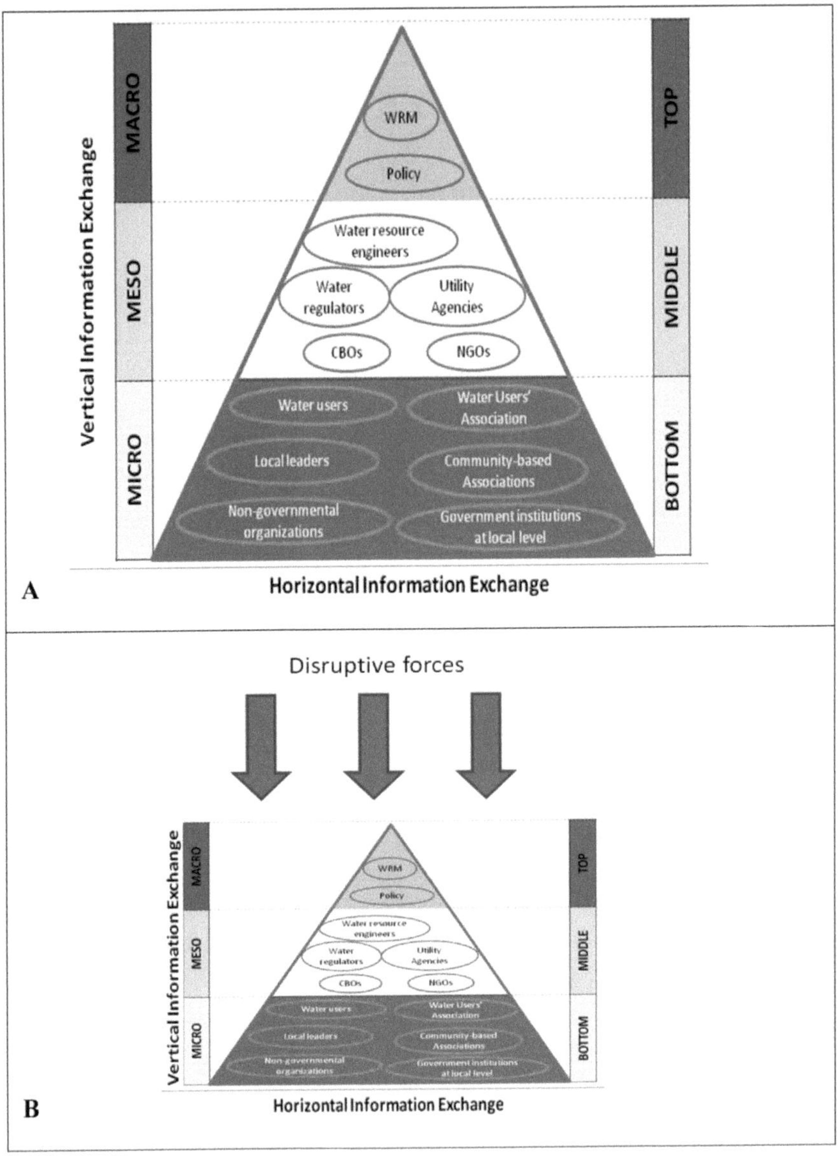

FIGURE 1.4 (See color insert.) The potential action of disruptive forces on the vertical information exchange: before (A), after (B).

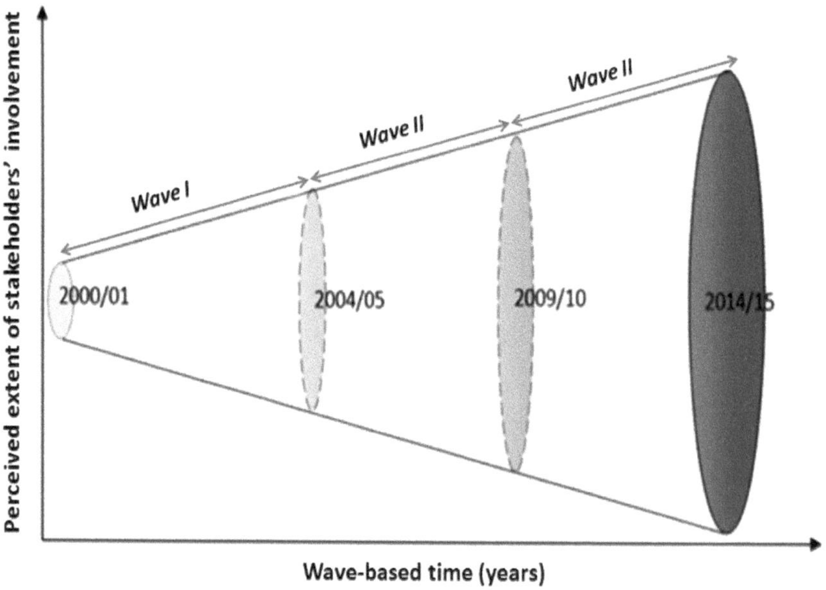

FIGURE 1.5 (See color insert.) Trends and aggregation of disruptive forces during different waves.

Source: Adapted with permission from Ref. [32]

Mobile devices including laptops, mobile phones, and personal digital assistants (PDAs) as well as the improved mobile networks created an environment for ICT to support a wide spectrum of stakeholders in the water basin management. This transformative power is attributed to the fact that mobile devices including phones have no specific social sphere.[36] Alongside the enhanced portability, there was also an increase in the processing power and storage capacity, which improved the desktop-based solutions. For example, the third wave from 2010/11 to 2014/15 was characterized by relatively more advanced and accessible computing power, mobile hardware, and networks. Computing power continued to increase while portable and fast hardware that supports big data storage increasingly became affordable. Software including mobile applications and open source also became accessible and networks coverage and costs became more inclusive. All these were opportunities to advance solutions by communities themselves.

1.4.3 POTENTIAL TRANSFORMATIVE POWER

Transformation takes place when activities contribute positively to achieving a social goal. Understanding the transformative power of mobile phone was enhanced through correlating the selected activities, tools, and goals. Results were validated using the effect of sizes. Effect sizes for interpreting the practical significance of correlation coefficients have been suggested by a number of authors[16,23] in statistics and behavioral studies. They suggested effect sizes' rule of thumb with intervals and five levels from negligible (±0.00 to ±30), low (±30 to ±50), medium (±0.50 to ±0.70), high (±0.70 to ±0.90) to very high (±0.90 to ±1.00). However, the most common one is Cohen's rule of thumb.[5,6] It suggests three levels of interpretation of Spearman's correlation from small (± 0.10), medium (± 0.30), to large (± 0.50). Based on the three levels of relevance (low, medium, and high) adopted in this study, the Cohen's rule is used with slight modification (Table 1.3).

TABLE 1.3 Rule of Thumb for Interpretation of the Effect Sizes in Correlation Analysis.

Positive correlation		Negative correlation	
Size	**Interpretation**	**Size**	**Interpretation**
0.70–1.00	High	−0.70 to −1.00	High
0.50–0.70	Moderate	−0.50 to −0.70	Moderate
0.30–0.50	Low	−0.30 to −0.50	Low
0.00–0.30	Negligible	−0.00 to −0.30	Negligible

Source: Adapted with slight modification from Ref. [5].

High and very high levels are merged to high for the purpose of confirming the context requirements. In this case, associations with Spearman's correlation, less than ±0.30 were considered negligible. Table 1.4 shows the summarized findings of the extent to which the various mobile phone services can bring transformation to the social, economic, and ecological dimensions.

1.4.4 SOCIAL TRANSFORMATION

Access to and use of mobile phones can create a greater social cohesion among FFO members, thereby leading to improved social relationships.

This is supported by a study on the use of mobile phones among farmers for agricultural development.[3]

TABLE 1.4 Mobile Tools as a Bridge Between Activities and Sustainability.

Activity	Cohon's RoT	Mobile phone tool	Cohon's RoT	Sustainability indicator
Decision-making	Medium	Short messages	Low	Water sanitation
Decision-making	Medium	Image	Medium	Controlled GHG
Decision-making	Medium	Video	Low	Encroachment of water sources
Planning	Low	Voice	Low	Human population
Policy-making	Low	Voice	Medium	Human population
Law enforcement	Low	Voice	Low	Water sanitation
Infrastructure maintenance	Low	Voice	Low	Water sanitation
Capacity building	Low	Short messages	Low	Human population

Farmers are "silent" researchers or in a more formal name "citizen scientists." They often collaborate with mainstream researchers/scientists in planning and implementing field research activities. The mobile phone has been a prominent research planning and implementation tool. The mobile phone-supported social networks between researchers and farmers in developing countries have a great potential to grow. SMS, image, and video services were perceived to offer medium potential in supporting the decision making during the time of the study. The mobile phone supports data collection that informs the decision making. Given the population of farming communities in the developing world, farmer-generated data far bypasses many manual operations and therefore contributes to overwhelming levels of volume data. While the linkage between the tools and goals seems to be low, there exists a room for improvement and therefore contribute more to transformations in the rural settings.

Earlier findings on this relationship in other sectors[7,9,20,21] showed that for a greater impact in the rural area, a combination of community radio and a mobile phone would be ideal in promoting engagement at BoP. Radio, a traditional ICT tool, has been important in informing especially in the low-income brackets of water resource stakeholders. Thanks to the convergence of ICTs, the mobile phone is providing audio services by accessing the various FM radio channels. This calls for more research related to the usage of the mobile phone in supporting activities that are aligned to achieving the goal, hence delivering the envisaged value. This implies that any designed mobile supported information system that is likely to make these tools converge would increase their association with activities they support at the BoP of water resource use communities.

1.4.5 ECONOMIC TRANSFORMATION

FFO implements economic activities like planning, banking, and financial mobilization. The traditional banking and financial mobilization systems do not favor much the rural areas where investment does not seem feasible. Mobile phones have come to the rescue where it complements other rural microfinance facilities as the best alternative to traditional banking. Mobile banking transforms the way rural dwellers keeps and transfer them money. A farmer group in the formalized irrigation association collectively makes decisions with regards to how much and from which source of water to be used in the production process. This is the economic decision with regards to usage of the scarce resource.

Will Marler[28] observed that apart from supporting individual activities, the mobile phone had a role to play in improving coordination of a set of activities, thereby reducing digital inequality between rural and urban areas. The video, image, and SMS tools were perceived to support these activities at a medium level according to validated Cohon's RoT. Voice messages were perceived to contribute to policy-making and planning at a low Cohen's RoT, while other tools had nonsignificant contributions (Table 1.4). However, on achieving the goal, which is the value FFO seeks to gain, voice message-supported policymaking had a medium contribution to achieving a goal indicated by the proportion of the human population in demand for water resources.

1.4.6 ECOLOGICAL TRANSFORMATION

Ecological sustainability of water resources is related to continuity in production in the irrigated agricultural lands. The protection of water resources against the factors causing the degradations is, therefore, an important aspect for FFO. Activities such as decision making, enforcing the laws, infrastructure maintenance and capacity building have been considered among interventions to achieve ecological sustainability. Except for capacity building where SMS contributes at low level, the rest had voice messages contributing the same level. This calls for specific measurements to ensure that mobile phone services contribute its potential to achieve the ecological transformation much needed in the water sector especially amid the changing climate. Global debates on environment and mobile phone have skewed toward the negative sides, such as its e-waste mismanagement and its carbon footprint.[36] Despite this trend, the positive side of the mobile phone as an ecological transformative tool worth heightened advocacy in developing countries.

1.5 SUMMARY

This chapter focuses on the disruptive and transformative power of mobile phones in the LVB. The background about the two concepts as well as the principle that guides their understanding are presented in brief. The contemporary debates among optimists and pessimists of the disruptive and transformative power of mobile phones in developing countries are summarized to inform the contextual zoom for the FFO in the LVB of Tanzania. Employing the activity–goal–approach, the potential of mobile phone tools to contribute to disrupting and transforming small-scale FFO are narrated and supported by empirical findings. Disruptive powers of mobile phones are presented and discussed within the ability to change the hierarchical, bureaucratic, and vertical communication structure into a more common communication platform where the macro, meso, micro scales of water resource management can meet for dialogues and decision making. Transformative powers of the mobile phone are presented and discussed within the framework where the application of mobile phone tools changes the way activities are done to realize more value in terms of achieving social, economic, and ecological transformation. Despite the disruptive and transformative powers of a mobile phone, its potential in transforming rural areas especially FFO that utilize water resources has not been fully tapped. More detailed conclusive research is needed in the nexus of activities, tools, and goals to inform better value attainment for sustainable irrigated agricultural production in the LVB.

ACKNOWLEDGMENTS

The study was partially financed by German Academic Exchange Services (DAAD) through a grant to support In-country PhD scholars in Tanzania. The Inter-University Council for Eastern Africa (IUCEA) supported the other part through the Lake Victoria Research Programme (VicRes). The University of Dodoma provided in-kind support through a study leave offered to the first author.

KEYWORDS

- **disruptive technology**
- **Lake Victoria Basin**
- **mobile phone**
- **small-scale agriculture**
- **transformative technology**
- **water resource management**

REFERENCES

1. Aubert, J. E. *Promoting Innovation in Developing Countries: A Conceptual Framework*; Policy Research Working Paper, No. 3554; World Bank; Washington, DC, 2005, p 39.
2. Chen, P. *Electronic Engagement: A Guide for Public Sector Managers*. ANU E Press: Canberra, 2007, p 122.
3. Chhachhar, A. R.; Hassan, M. S. The Use of Mobile Phone among Farmers for Agriculture Development. *Int. J. Sci. Res.* **2013,** *2* (6), 95–98.
4. Cobb, C. G., Ed. *Making Sense of Agile Project Management: Balancing Control and Agility*. John Wiley & Sons: New York, 2011, p 264.
5. Cohen, J., Ed. *Statistical Power Analysis for the Behavioral Sciences*. Lawrence Erlbaum Associates (Routledge): Hillsdale, NJ, 1988; p 400.
6. Cohen, J. Things I Have Learned (so far). *Am. Psychol.* **1990,** *45* (12), 1304–1312.
7. Crandall, A.; Otieno, A.; Mutuku, L.; Colaço, J.; Grosskurth, J.; Otieno, P. *Mobile Phone Usage at the Kenyan Base of the Pyramid,* 2012; p 59. https://blogs.worldbank.org/ic4d/files/ic4d/mobile_phone_usage_kenyan_base_pyramid.pdf (accessed Jan 12, 2018).
8. Dodson, L. L.; Sterling, S.; Bennett, J. K. In *Considering Failure: Eight Years of ITID Research*, Fifth International Conference on Information and Communication Technologies and Development; ACM, New York, 2012; pp 56–64.

9. Duncombe, R.; Boateng, R. Mobile Phones and Financial Services in Developing Countries: A Review of Concepts, Methods, Issues, Evidence and Future Research Directions. *Third World Q.* **2009,** *30* (7), 1237–1258.
10. Evans, N. D. *Business Innovation and Disruptive Technology*. Prentice Hall: Upper Saddle River, NJ, 2003; p 173.
11. Mashable. *The 15 Countries with the Highest Smartphone Penetration*, 2013. https://mashable.com/2013/08/27/global-smartphone-penetration/#HPb4V3vekaqB (accessed Mar 24, 2018).
12. Harriden, K. Big Water Needs "Little" People: Improving Water Resource Management by Including Households. *Int. J. Water* **2014,** *8* (2), 111–127.
13. Heeks, R. Do Information and Communication Technologies (Icts) Contribute to Development? *J. Int. Dev.* **2010,** *22*(5), 625–640.
14. Heeks, R. Development 2.0: The ITenabled Transformation of International Development. *Comm. ACM* **2010,** *53* (4), 22–24.
15. Heeks, R. *Most e-Government-for-development Projects Fail: How Can Risks be Reduced?;* 2003; p 19. http://unpan1.un.org/intradoc/groups/public/documents/NISPAcee/UNPAN015488.pdf (accessed Mar 10, 2018).
16. Hinkle, D. E.; Wiersma, W.; Jurs, S. G. *Applied Statistics for the Behavioral Sciences*. Cengage Learning: Boston, USA, 2003; p 756.
17. Ibrahim, H. Technology Acceptance Model: Extension to Sports Consumption. *Procedia Eng.* **2014,** *69*, 1534–1540.
18. ITU: *ICT Facts & Figures,* 2015, p 6. https://www.itu.int/en/ITU-D/Statistics/Documents/facts/ICTFactsFigures2015.pdf (accessed Mar 16, 2018).
19. Jensen, R. The Digital Provide: Information (Technology), Market Performance and Welfare in Southern Indian Fisheries Sector. *Quarterly J. Econ.* **2007,** *CXXII* (3), 879–924.
20. Karanasios, S. Framing ICT4D Research Using Activity Theory: A Match Between the ICT4D Field and Theory? *Information Technol. Int. Dev.* **2014,** *10* (2), 1–17.
21. Heeks, R.; Ospina, A,. Eds. *ICTs, Climate Change and Development: Themes and Strategic Actions*. Centre for Development Informatics, Institute for Development Policy and Management: Manchester, 2012; p 431.
22. Khazaei Pool, J.; Verij Kazemi, R.; Amani, M.; Kia Lashaki, J. An Extension of the Technology Acceptance Model for the e-repurchasing of Sports Match Tickets. *Int. J. Manag. Bus. Res.* **2016,** *6* (1), 1–12.
23. Kotrlik, J. W.; Williams, H. A.; Jabor, M. K. Reporting and Interpreting Effect Size in Quantitative Agricultural Education Research. *J. Agric. Educ.* **2011,** *52* (1), 132–142.
24. Levi, C.; Kyazze, B. F.; Sseguya, H. *Effectiveness of Information and Communication Technologies in the Dissemination of Agricultural Information to Smallholder Farmers in Kilosa District*, 2014; p 4. http://www.ruforum.org/sites/default/files/Levi.pdf (accessed Mar 4, 2018).
25. Liu, I. F.; Chen, M. C.; Sun, Y. S.; Wible, D.; Kuo, C. H. Extending the TAM Model to Explore the Factors That Affect Intention to Use an Online Learning Community. *Comput. Education* **2010,** *54* (2), 600–610.
26. Lugo, C., Jordan, A.; Benson, D. The Role of Problem and Process Factors in Creating Effective Trans-boundary Water Regimes: The Case of the Lake Victoria Basin, East Africa. *Int. J. Water* **2014,** *8* (2), 219–240.

27. Maldonado, E. A.; Maitland, C. F.; Tapia, A. H. Collaborative Systems Development in Disaster Relief: The Impact of Multi-level Governance. *Information Syst. Frontiers* **2010,** *12* (1), 9–27.
28. New Media & Society. *Mobile Phones and Inequality: Findings, Trends, and Future Directions*, 2018. http://journals.sagepub.com/doi/abs/10.1177/1461444818765154 (accessed Apr 9, 2018).
29. Mihály C. Mobile Devices in Agriculture: Attracting New Audiences or Serving the Tech-savvy? *J. Agric. Informatics* **2015,** *6* (3), 75–84.
30. Millett, L. I.; Estrin, D. L., Eds. *Computing Research for Sustainability*. National Academies Press: Washington, D.C, 2012, p 156.
31. Molnar, J.; Matuha, M.; Atukunda, G.; Walakira, J.; Terhune, J.; Bukenya, J.; Naigaga S. The Role of Mobile Phones in Facilitating Aquaculture Development in Uganda. *World Aquaculture* **2016,** *47* (1), 39–46.
32. Mongi, H.; Meinhardt, M. Integrated ICTs for Water Basins Management in Southern Africa: Systematic Review and Meta-analyses for Perceived Relevance Criteria. *Agris on-line Papers in Economics and Informatics* 2016, *8* (2), 103–110.
33. National Consumer Council (NCC). *Deliberative Public Engagement: Nine Principles,* 2008, pp 18. https://www.involve.org.uk/wp-content/uploads/2011/03/Deliberative-public-engagement-nine-principles.pdf (accessed Jan 13, 2018).
34. Newzoo's: *Global Mobile Market Report*, 2017. https://newzoo.com/insights/rankings/top-50-countries-by-smartphone-penetration-and-users/ (accessed Mar 23, 2018).
35. Perlman, L. Technology Inequality: Opportunities and Challenges for Mobile Financial Services. Ph.D. Thesis, Center for Financial Inclusion at ACCION, 2017, pp 52. http://www.centerforfinancialinclusion.org/storage/CFI_MFS_Tech_FINAL.pdf (accessed Mar 26, 2018).
36. RESET, *The Social and Environmental Impact of Mobile Phones,* 2017. https://en.reset.org/knowledge/ecological-impact-mobile-phones (accessed Apr 12, 2018).
37. Salim, A. *Impact of Higher Learning Institutions Expansion on the Adequacy of Network Infrastructure in Developing Countries: A Case of ICT Planning at the University of Dodoma*, 2016, pp 65. http://repository.out.ac.tz/style/images/fileicons/application_pdf.png (accessed Mar 18, 2018).
38. Scientific Expert Group on Climate Change (SEG). *Confronting Climate Change: Avoiding the Unmanageable and Managing the Unavoidablel* Report prepared for the United Nations Commission on Sustainable Development, Washington, DC: Sigma Xi, Research Triangle Park, and the United Nations Foundation, 2007; p 144.
39. Selke, S., Ed. Lifelogging—Disruptive Technology and Cultural Transformation—The Impact of a Societal Phenomenon (Chapter 1). In *Lifelogging: Digital Self-tracking and Lifelogging—Between Disruptive Technology and Cultural Transformation*; Springer Fachmedien Wiesbaden: Germany, 2016; p 1–24.
40. Sife, A. S. The Contribution of Mobile Telephony, Radio and Television to Rural Livelihoods and Poverty Reduction in Morogoro Region, Tanzania. Ph.D. Thesis, University of Dar es Salaam, Tanzania, 2010; p 356.
41. Steyn, J. A Critique of the Claims about Mobile Phones and Kerala Fisherman: The Importance of the Context of Complex Social Systems. *Elect. J. Information Syst. Dev. Countries* **2016,** *74* (3), 1–31.

42. Stouthuysen, P.; Le Roy, D. *Cradle to Cradle: Theoretical Framework*, 2010; p 28. http://www.c2c-center.com/library-item/cradle-cradle-theoretical-framework (accessed Feb 13, 2018).
43. The World Bank, Data: Population, 2018. https://data.worldbank.org/indicator/SP.POP.TOTL (accessed Mar 23, 2018).
44. United Republic of Tanzania (URT). *National Water Policy*, 2002, Dar es Salaam: Ministry of Water; p 49.
45. Urwin, T., Ed. *ICT4D: Information and Communication Technology for Development*. Cambridge University Press: Cambridge, 2009; p 404.
46. van den Hoof, B.; de Ridder, J.; Aukema, E. Exploring the Eagerness to Share Knowledge: The Role of Social Capital and ICT in Knowledge Sharing', In *Social Capital and Information Technology;* Wulf, H., Ed.; Cambridge: MIT Press, 2004, p 163–186.
47. Weiss, C. H. How Can Theory-based Evaluations Make Greater Headway? *Evaluation Rev.* **1997,** *21*, 501–524.
48. Wilson, J., Ed. *Overcoming Gender Inequalities through Technology Integration: Advances in Human and Social Aspects of Technology*; IGI Global: Hershey, PA, 2016; p 324.
49. World Bank. *The Little Green Data Book*. International Bank for Reconstruction and Development/World Bank: Washington, DC, 2014; p 248.
50. World Economic Forum. *4 Reasons Developing Countries Need Digital,* 2015. https://www.weforum.org/agenda/2015/03/4-reasons-developing-countries-need-digital/ (accessed Jan 08, 2018).

CHAPTER 2

Transformative Power of Mobile Phones in the Indian Agro-Supply Chain

GEORGE JOSEPH and K. A. ZAKKARIYA

ABSTRACT

This chapter focuses on the impact of mobile technology on agricultural productivity in India through services for small-scale farmers and agricultural businesses. Information requirements in an agriculture supply chain can be categorized as follows: (1) know-how: fundamental information about crop selection and seeds; (2) contextual information: weather and best practices to be adopted; (3) market information: prices and demand indicators; and (4) logistical information: access to market and mode of transportation. Mobile products and services have shown their ability to deliver real-time information directly to the farmer to improve traceability, quality control, and logistics systems and to enable agricultural businesses to operate more efficiently. Mobile financial services are enabling farmers to receive and make payments and access financial services. At each stage of the agricultural value chain, mobile phones can help to improve communication and efficiency, supporting increased food production and better livelihoods. Use of mobile phone is improving data visibility and supply chain efficiency, that is, optimizing supply chain management across the sector and delivering efficiency improvements for transportation logistics, including through tracking systems and mobile management of supply and distribution networks. There are many constraints and challenges in the form of inadequate information and issues of converting content into local language, character limits for SMS services, handset compatibility, illiteracy, and so on. There are also many instances of new practices in mobile-phone-based information delivery characterized by its real-time nature, which improves traceability, quality control, and access to financial services. While exploring the potential of mobile agriculture services, four areas are identified in this chapter. These areas are as follows: Improved access to financial services tailored for agriculture-mobile payments, insurance, and lending platforms. Another area that has shown progress is agricultural information like delivering information relevant to farmers (commodity prices and weather forecasts and advice on agricultural techniques), improved data visibility and supply chain efficiency: mobile management of supply and distribution networks, tracking systems, improvement in logistics, and so on. The chapter will also look into enhancing access to the market through mobile trading and tendering platforms by linking traders, buyers with farmers.

2.1 INTRODUCTION

Indian farmers form about half of the Indian population and generate just one-fifth of gross domestic product and they stay mostly deprived and untrained. India has extensive research and development facilities in the agriculture sector. Information relevant to farmers from these centers needs to reach the required places. Information dissemination influences the growth of agriculture worldwide. Mobile phones have penetrated all around the country as the cost of ownership and call charges have come down significantly. Mobile-phone-based information delivery to farmers can prevail over information asymmetry in agro-supply chain. Mobile-phone-assisted agricultural services have an impact on farmers and the supply-chain efficiency.

Therefore, great scope exists for greater rural productivity by bringing the mobile-phone-based information delivery as a part of the farmer's supply chain. To fight this state of affairs and take advantage of the fact that the 1.3 billion population has approximately one billion mobile subscriptions, several smart agricultural apps that deliver data such as moisture content of soil, mineral levels, prices, and so on to mobile phones are slowly gaining popularity. These inventions epitomize an accelerating trend in which firms are utilizing mobile penetration to help farmers progress in several dimensions and develop bottom lines. For example, Jayalaxmi Agro Tech developed an app that gives uneducated farmers with sufficient crop-related info[9] using audiovisual tools. Another one is called Mandi Trades catalog that updates essential crop price from the government to distantly located farmers. There is Rainbow Agri for neighborhood buyers and sellers. MPower Social app provides veterinary guidance for cattle owners.[7] Thus, these new breeds of mobile apps are in the process of revolutionizing life for farmers in the country and abroad.

This chapter focuses on the transformation that is happening in the agro-supply chain by the use of mobile phones.

2.2 METHODOLOGY

The study is based on a nonsystematic review of various resources collected from various agri-tech websites, case studies, and so on. The study tries to answer the following questions: What are the challenges faced by Indian agriculture before and after the advent of ICT? What are the requirements and challenges faced presently by the various agri-tech firms present in the Indian market? What are the opportunities and future of this sector?

The study thus includes a critical description of findings from a nonsystematic review of the collected set of 46 sources.

2.3 CHALLENGES

Although lots of problems like connectivity, cost, basic literacy, and computer literacy hold back the speedy growth of mobile agriculture, yet it will certainly be a mechanism of growth in rural India once the problems are solved to an extent. The reach of agricultural technology is still meager and needs to cover a larger farmer base. The delivery and spread of technologies are not homogeneous throughout the nation.[17] The use of technology is prolific among the already rich farmers and by utilizing these services they are further prospering.

There is a need to ensure the quality, timeliness, and trustworthiness of the information. The dynamism in the agricultural practices and the responsiveness to the market holds the key in the context of rising food prices and growing population where the concern of food security rises. Improvement in productivity depends on research, extension, literacy, and infrastructure. Information and communication technology plays a vital role in the knowledge transfer based on geographical relevance, cropping system, good agricultural practices concerning input requirements, and timely use of fertilizers. The supply chain uncertainties caused by demand variability, geographical variability of prices, weather, new variety of seeds, and so on make the traditional knowledge of farmers powerless.

Extension services though available tend to break down very often and have the limitation of not reaching all the farmers, at the right cost and time. This is where cheap mobile information services become an appropriate channel to connect with farmers.

Major reforms have been brought to Indian extension services with incorporation of information communication technologies improving efficiency, effectiveness, and timeliness of information being sent. This became possible due to the proactive response of private entities in the agriculture sector like Reuters market light, Nokia life tools,[3] and so on. We have thus witnessed technical innovations, increase in information penetration, and increase in overall supply chain efficiencies. The content to be transmitted is formulated, customized in local language, and made available to farmers on their mobile phones. This system facilitates two-way communications through SMS, toll-free call centers, GUIs, and so on. The agricultural supply chain is constrained by slower financial flows due to the lack of collaterals to avail loans.

The weaknesses are due to inadequate market information coupled with hindering policy, legal and regulatory framework, and low-income levels among farmers and consequently investment in tools for yield is unthinkable or delayed. Monetary transactions in most of the cases require a location and transportation, which can be problematic particularly in the rural areas of India.[30] Penetration of mobile-based payment systems in the rural areas will bring positive impacts by overcoming the infrastructure and the time constraints. Such systems can increase transparency, efficiency, and reduce transaction costs. Therefore, there is a need for integrating financial supply chain with the physical supply chain. There is also an urgent need to ramp up technology-based delivery channels in India.[31]

2.4 INFORMATION

Modern agricultural supply chain supports three kinds of network flows simultaneously[6]:

- Physical: product movement suppliers to final customers.
- Financial: loan requirements, repayments, savings and insurance, and so on.
- Informational: facilitating and coordinating physical and financial flows.

All these three need to be regular and uninterrupted for the success of the supply chain. Physical product flow occurs due to the farmer's necessity of selling the product or as a result of increasing demand. Government has taken many steps such as rural road connectivity and collection centers to ensure physical flow. But lack of information and financial flows make the product flow inefficient and sometimes uneconomical. Mobile-phone-based information delivery is characterized by its real-time nature, which improves traceability, quality control, and accesses financial services. While exploring the potential of mobile agriculture services, following four areas have been identified[44]:

- Improving access to financial services: financial services tailored for agriculture purposes such as mobile payments, insurance, and lending platforms.
- Agricultural information: delivering information relevant to farmers (commodity prices and weather forecasts and advice on agricultural techniques.
- Improving data visibility and supply chain efficiency: mobile management of supply and distribution networks, tracking systems, improvement in logistics.
- Enhancing access to market: through mobile trading and tendering platforms.

Therefore, smartphone-enabled services are catering to the needs of entire gamut of farmer.[26] The real payback from information depends and alter based on region, crop, infrastructure, and socioeconomic conditions but will definitely enable farmers to make informed decision rather than the present practice of guesswork on time of sowing, application of fertilizers, irrigation, and so on. The market information, on the other hand, helps farmers to get connected to input market and help over expenditures. Thus, efficient mobile agriculture services can realize the much-needed financial inclusion.[25]

2.5 ECOSYSTEM

Multistakeholder partnerships can bring institutional expertise into the agri-supply chain. Thus, cooperative models are required at the field level and local subsidiaries are essential for offering value-added services.

Institutional learning programs can create changes and improvement among farmers and value chain players. Strong brands and positioning of mobile agriculture players can help the development of the ecosystem. It is required to make available minimum hardware and software applications even beyond the mobile phones along with other infrastructure to speed up the implementation.[33]

The culture, sensitivity, and acceptance of technology are intertwined and differ from location to location. There must be information agents who can put in time and effort to train and empower others. The younger generation can take up this role as they are fast learners and are actual intermediaries between adults and technology. They can share new knowledge through informal approaches of family learning. All these can start with a mindset of "learning how to learn" in which farming community learns to solve problems by themselves.

Team-building model forms the basis for such collaborative networks where facilitators work closely with farming communities.[35] This model is scalable with more area and networks in which partners having varying levels of competence and domain expertise, technology know how, and familiarity will be able to engage and track the progress of the community usage and manage data.[35] Therefore, a higher degree of support in the initial phases of a partnership is essential for the success of this model. For instance, in a new area of work, there is a need to devote time and effort to necessary technical and process-related support for local human resources. This can be achieved through learning, experimenting, and iterating the approach. The lessons learned and learning curve concerns the measurement of social impact too. Indirect methods like health and nutrition can have a better reach which can bring ripple effects of good will.[35] Thus, indirect approaches instead of solely targeting beneficiaries of the community along with efforts to reach those who can adopt a practice, and who can act as promoters and influencers will bring in greater results.

2.6 POLICIES

The National Policy for Farmers (2007) was formulated and approved by the Government of India based on the recommendations of National Commission on Farmers that was formed in 2004 and the suggestions received from Central Ministries and Departments and Center and state governments. It has important provisions for the use of technology to enhance productivity aimed at a revolution capable of improving productivity in a sustainable manner. Also, the National Telecom Policy (2012) was formulated to improve broadband penetration by which mobile phones can become an instrument of socioeconomic growth and empowerment.[20]

National Mission on Agricultural Extension and Technology aims to rebuild agricultural extension services in a way to deliver appropriate technology and practices to farmers. This requires a fine blend of outreach, dissemination, use of ICT, popularization, and reinforcement to encourage the use of machines, accessibility to quality seeds, fertilizers, pesticide, and so on. Clusters of Interest Groups and Farmer Producer Organizations are getting formed.[8] National optical fiber network targets range of services in rural India. Mobile value-added services for rural women's self-help group (SHG) is another noteworthy initiative. In this regard, the SHGs are made available with information in local languages through mobile phones through SMS, integrated voice response system, and so on. Bharat Nirman Kendra provides feedback on the quality of implementation of the program.[2]

2.7 APPLICATIONS AND SERVICES

Types of services that can be offered through mobile to improve supply chain efficiency and productivity can be classified as follows:

- Enabling access to local markets: enables small-scale producers to work along with cooperatives through an e-commerce platform.
- Field audits: use tablets to improve efficiency for auditors monitoring quality.
- Information services: provide tailored agriculture information over mobile phones to the farmers.
- Payments and loans: integrates farmers into a registered database and enables access to loans and payments via mobile money. Smartphone-enabled market place with payments, loans, insurance, and so on is envisaged.[32]
- Receipt services: provides recordkeeping services, mobile registration, and receipts.

Modern ICT models in India for information delivery to farmers are as follows:

- Internet based—knowledge centers, ITC echoupal, E-Sagu, and so on.
- Mobile app—fisher friend-MSSRF, Nokia life tools, Tata MKrishi, and so on.
- Mobile SMS—Reuters Market light, Warna unwired-Microsoft, KVK's—NAIP, Kisan Sanchar, and so on.
- Mobile voice—IFFCO Kisan Sanchar Limited, and so on.
- Tele-center based—KISAN call center, BSNL helpline, and so on.
- Video based—digital green.

The MKrishi platform provides customized services in local languages via mobile phones.[37] Such services enable farmers remotely get connected with stakeholders, thereby giving accessibility to good-quality seeds, fertilizers, and other inputs, along with data-driven advices on agricultural practices, including market information like prices, weather, and so on.[16] Kisan credit card provides affordable, timely, and adequate credit for farmers in India.[13] At Kisan choupal, discussions happen around problems and solutions assisted by IT. Technical videos, movies, and so on, are used for these purposes.[14] Kisan SMS Portal provides SMS messages giving information, delivering service and advisories from specialists, scientists, and officers. Irrigate via smartphone aims at monitoring and controlling crop irrigation systems. Moisture sensors can give precise management of all inputs like fertilizers.[19]

GPS mapping for an input using variable rate technology helps a farmer to access the need of the soil at various locations. This technology can be extended to documentation about yield, moisture, maps, and so on.[36] There are also cams that observe crops and take the scientists' expertise while they being off the field.[24] Farming software algorithms can now track of crops, predict yields, time to plant, what to plant, to intercrop or not or determine the current crops needs. These can be done more effectively with community involvement in agriculture. Site-specific crop management has enabled observation, measurement, and response to inter- and intrafield crops requirements. These techniques have optimized resource usage like fertilizer, pesticides, water, and so on while improving quality and quantity, thus reducing the cost of production preventing soil degradation.[21]

2.8 REQUIREMENTS

For a successful acceptance of innovation by the target market, there are two requirements:

- Innovation needs to be scientific in terms of technology and core competencies, and
- innovation must be supported by business operational strategy focusing on sustainability and commercialization.[10]

Brand building can inspire trust among farmers and this can be achieved through quality and consistency of mobile services. It is also very essential to build partnerships with input and output companies in this sector.[1] Dedicated field staffs who can establish trust and goodwill are essential. A long-term strategic approach by firms without short-term profit objectives is vital.

- Firms must associate with the various national and international organizations for procurement, banking, insurance, quality certification, and access to international markets, and so on.[12]
- Local accessibility and prompt aftersales service for 3–5 years can help sustain and scale business.[43] Demonstrations and pilot installations are the way forward.
- New solid and dominant business model needs to emerge which can be followed across the country. Business inclusion models must involve a well-developed strategic collaboration strategy among critical stakeholders.
- Personalized content like that provided by MKrishi enables personalized farming through an integrated cross-platform delivery system.
- Realization of the positive socioeconomic change through technology can build trust. Once farmers experience practical savings and profits, they will be more optimistic about the technology.[34]

2.9 CHALLENGES FACED BY VARIOUS MOBILE APP PROVIDERS

- A national agenda for precision agriculture is still not very clear. Scaling up platforms across locations while developing modular localization and optimization is still far away. Market-driven production approaches are yet to get popularity. This will provide traceability and transparency in the value chain comparable to industrial production.[22] Implementing crop insurance, better storage facilities, and processing units can help better control of the processes.

- Aging prevents farmers from learning new technology and its application.[40] The issue is aggravated when youth still leave the agricultural community for greener pastors in urban areas.
- Agri-output is seasonal and revenue declines in some seasons creating additional challenges in demand for such products and continuous subscriptions. Unpredictable natural disasters policies can add to the woes.
- Competition after the technology is accepted leads to a setback for the pioneers who have painstakingly devoted their times for building a market cannot be underestimated.
- Culture and perceptions of local people vary from place to place therefore demands different learning curve for technology acceptance.
- Difficult terrain prevents field engineers from reaching ICT installation at odd hours.
- Field trials still require long permission procedures to convince the government. Farmers are reluctant to pay for new service, hence there is a constant need to reduce the subscription fees.[38]
- Getting financial support for less profit-intensive business models in this domain.
- Global warming and climate change present new challenges and unpredictability, which can be a blessing in disguise with better technologies.
- Government policy changes in each term and conflicts of interest between center and states.
- Illiterate farmers and stubborn conventional mindsets prevent change to modern technology. Confidence building among farmers requires patience.
- Lack of sustainable business model: The competition among the new ICT players are fragmenting the market and again disintegrating the supply chain.[5]
- Lack of tried-and-tested research outcomes on the implementation of ICT is a challenge. We have a few models emerging anyway.

- Mobile incompatibility as an issue is getting reduced but network connectivity issues still remain and cause issues when mobile-based agro-advisory integrates several platforms through technologies. Communication to servers and sensors at the field are dependent on these. Weather station and sensors are vulnerable and extreme conditions like monsoons can affect charging via solar panels.
- Resistance of technicians to work in the poorly facilitated rural sector.
- Retaliation from middlemen in various forms.
- Significant funds are to be allocated for marketing, advertising, promotion, and training from a long-term point of view. Thus, developing a sustainable business model to convince investors is the real challenge. Inertia in the rural market is very high and this does not go well with investors' financial calculations. Huge difference in the cost incurred by R&D, promotion campaigns with the selling price to farmers extends the break-even time of these technology products.[11]
- Social issues like caste system, which is still rooted in rural India.

2.10 OPPORTUNITIES

- A to-and-fro communication enabling the formulation of government policies and feedback thereafter between farmers and government can help develop further policies.
- Advice on pesticides and fertilizers such as how much and when to spray, when to harvest in relation to whether to limit crop damage are much sought after.
- Agro-product companies having direct access to farmers can collaborate with mobile app service providers and take the financial burden of subscription fee subsidized for the entire network.
- Banks and insurance companies can use the mobile app service providers who can provide pertinent information relating to loans, insurance, and so on.[28]
- Combining yields allows efficient collection of goods and help in managing market prices so that they can choose where and when to sell.
- Designing usage-based revenue models and low-cost transactional platform are the need of the hour in agri-transformation.[27]

- Flexibility in accommodating regional characteristics and integration into existing systems of various stakeholders are the challenges that need to be fully dealt with to make the next big move.
- Holistic approach brought in by ICT in agriculture through environ-friendly techniques such as biological control, integrated pest management, and optimal use of nutrients helps planning for environmental factors. It also enables tracking the organic stature of a farm.[4]
- Increase in the disposable income of farmers will lead to demand-led diversification and better revenues.[29] ICT-enabled collective bargaining can also lead to better prices and lower input costs.[18]
- IT-enabled services create direct employment opportunities in rural areas through round-the-year farming activities and secondary employment in logistics, packaging, and other supply chain segments.[42]
- Many firms are coming into the market to provide cheaper access to credit thereby liberating farmers from the hands of moneylenders.[15] Risk mitigation plans provided by technology platform protects farmers from losses, suicides, and exoduses.
- Mobile-based services are also foraying into health care, bringing facilities to the doorstep of the member farmers.
- Mobile-phone-enabled services are processing huge amount of data pertaining to local weather, yield, and farming practices.[39] This can create new opportunities for newer and better uses of data collected.
- Such endeavors provide the firm with new markets while directly contributing to an existing business, branding, and public relations.
- The impacts of technology have already prompted more institutions to implement it in their purviews. Such validations and news can spread throughout the agri-industries, which in the future can shape government policies toward digitizing farm data. This can bring the next revolution in agriculture, food processing, and other allied industries.

2.11 FUTURE PROSPECTS

Working with local community groups such as self-help groups, women's groups and farmers' groups need a fresh approach to achieve still better results in villages. Working with strategic partners, sharing content, and mobilizing community group while building grassroots-level institutions requires training for trainers in social mobilization. Exploring ways to leverage existing community groups of students, youth, and women as the logical sources of partnerships need to be done.[41] Online learning platforms are to be developed for extension officers that can help to deliver better quality training supported by a curriculum that supports experiential learning focusing on technical skills like farming know-how and process-related skills such as content creation, dissemination, and data monitoring. With on and offline components, usage data and feedback, the quality of future-training programs can be improved on a continuous basis. These hybrid massive online open courses for mediators help personal facilitation and assessment.[23]

2.12 SUMMARY

This chapter focuses on the transformation that is happening in the agro-supply chain by the use of mobile phones. The challenges in this sector are too many and IT had too much to solve in this domain. The data and information hold the key to unlock the potential of this sector and innovations and policies are the ways forward to develop a strong ecosystem. Mobile apps and services have already made their foray into this sector and have started influencing the lives of farmers positively. But there are requirements and challenges to be dealt with to exploit the opportunities even further. The future is bright and the transformation is nothing short of a phenomenon.

KEYWORDS

- **agricultural productivity**
- **agro-supply chain**
- **mobile agriculture services**
- **mobile technology**
- **transformation**

REFERENCES

1. Beaver, E. 2015. https://webdam.com/blog/how-to-build-a-brand-people-trust (accessed Feb 3, 2018).
2. Care Research, 2017. https://researchreports.careratings.com/industries/annual-review/industry-section-52-2013/universal-service-obligation-fund-usof-187.html (accessed Jan 30, 2018).
3. Casper Van. 2012. https://www.theguardian.com/global-development-professionals-network/2012/nov/27/farmers-mobile-phones-sms-agriculture (accessed Jan 2, 2018).
4. Chen, W. K.; Miao, C. H. 2013. https://www.ncbi.nlm.nih.gov/pmc/articles/PMC5620702/ (accessed Feb 12, 2018).
5. Danish Business Authority. 2012. https://www.oecd.org/innovation/inno/49537036.pdf (accessed Feb 12, 2018).
6. Das, A. 2016. https://www.mbaskool.com/business-articles/operations/5081-integration-the-way-forward-to-differentiate-agriculture-supply-chain.html (accessed Jan 20, 2018).
7. Debeshi, G. 2017. https://www.enterpriseinnovation.net/article/indias-farmers-turn-mobile-apps-1956219016 (accessed Jan 2, 2018).
8. Department of Agriculture and Cooperation, Government of India, 2017. http://vikaspedia.in/agriculture/policies-and-schemes/crops-related/krishi-unnati-yojana/national-mission-on-agricultural-extension-and-technology (accessed Jan 30, 2018).
9. Didem, T. 2016. https://modernfarmer.com/2016/01/agriculture-apps-india/ (accessed Jan 2, 2018).
10. Dintrans, P.; Amit, A.; Bhattacharya, J. 2014. https://www.cognizant.com/Insights Whitepapers/Output-and-Outcome-Based-Service-Delivery-and-Commercial-Models.pdf (accessed Feb 3, 2018).
11. E-Agriculture. 2013. http://www.e-agriculture.org/forumtopics/what-are-specific-constraints-you-have-faced-use-icts-resilience-december-2nd (accessed Feb 12, 2018).
12. Global Impact Investment Network. 2017. https://navigatingimpact.thegiin.org/strategy/improved-market-linkages/ (accessed Feb 3, 2018).
13. Hans India. 2015. https://m.dailyhunt.in/news/india/english/thehansindia-epaper-hans/transforming+agriculture+with+e+technology-newsid-67860935 (accessed Jan 30, 2018).
14. Hans India. 2015. http://www.thehansindia.com/posts/index/Young-Hans/1970-01-01/Transforming-agriculture-with-e-technology/301765 (accessed Feb 3, 2018).
15. Hardiker, J. 2006. http://www.indiatogether.org/vidlender-opinions (accessed Feb 12, 2018).
16. Inclusive Business Hub. 2013. http://www.inclusivebusinesshub.org/project/project-profile-mkrishi-mobile-technology-for-farmers-in-india/ (accessed Jan 30, 2018).
17. Insights. 2014. http://www.insightsonindia.com/2014/11/27/e-technology-aid-farmers/ (accessed Jan 2, 2018).
18. Kearney, J. 2010. https://www.ncbi.nlm.nih.gov/pmc/articles/PMC2935122/ (accessed Feb 12, 2018).
19. Kumar, R. 2015. http://krishisewa.com/articles/soil-fertility/600-tools-for-nitrogen-management.html (accessed Feb 3, 2018).

20. Lekshman, S. P. 2017. https://medium.com/@LuckyWrites/how-can-farming-and-technology-go-hand-in-hand-in-a-country-like-india-ad06d42f555b (accessed Jan 30, 2018).
21. Madhavan, N. 2012. https://www.businesstoday.in/magazine/special/innovation-agriculture-precision-farming/story/186624.html (accessed Feb 3, 2018).
22. Maras, E. 2016. https://www.foodlogistics.com/technology/article/12177254/fb-tackles-supply-chain-traceability-head-on (accessed Feb 12, 2018).
23. Mercado, V.; Martín, A.; Beltran, J. 2017. https://repositorio.itesm.mx/bitstream/handle/11285/621210/ConnectivityofLearningMOOCApaRepo.Pdf (accessed Feb 12, 2018).
24. MIleff, G. 2018. https://hubpages.com/food/The-Role-of-Information-Technology-in-Agriculture (accessed Feb 3, 2018).
25. Mittal, S.; Mehar, M. 2012. https://ageconsearch.umn.edu/bitstream/55478/2/2_Mittal.pdf (accessed Jan 20, 2018).
26. Mustaquim, M. 2015. https://www.ruralmarketing.in/industry/technology/mobile-services-can-increase-farmers-income-by-rs-56-000-cr (accessed Jan 20, 2018).
27. Mutabazi, P. 2018. https://www.linkedin.com/pulse/managing-innovation-m-pesamobile-money-success-case-study-mutabazi (accessed Feb 28, 2018).
28. Mymoneymantra. 2012. http://www.mymoneymantra.com/blog/smartphone-a-portable-bank-pocket/ (accessed Feb 12, 2018).
29. Okunmadewa, F. Enhancing Farm Gate Prices: The Role of Agricultural Marketing Research. *J. Econ. Rur. Dev.* **2008,** *16*(1), 27–33.
30. Pandey, T., Nagahari, V.; Venetia, M. Innovative Payment Solutions in Agricultural Value Chain as a Means for Greater Financial Inclusion. *Agric. Econ. Res. Rev.* **2010,** *23* (1), 527–534.
31. Pandey, T. 2010. https://ageconsearch.umn.edu/bitstream/96929/2/17-Tushar-Pandey.pdf (accessed Jan 20, 2018).
32. Rao, N. H. 2017. profile/N_H_Rao/publication/316430764_Research_to_Innovation Trends_in_Agricultural_Research_Management_in_India/links/58fdacf14585152eded4c597/ (accessed Jan 30, 2018).
33. Saiplatform. 2015. http://www.saiplatform.org/uploads/SAI_Platform_publications/SAI_Platform_Farmer_Partnership_-_Practitioners_Guide_-_May_2015.pdf (accessed Jan 20, 2018).
34. SMJR Consult. 2011. https://www.fsnnetwork.org/sites/default/files/a_guide_for_smallholder_farmers_on_saving_accessing_credit_and_effectively_managing_money_for_improved_livelihoods.pdf (accessed Feb 3, 2018).
35. Sylvester, G. 2015. http://businessdocbox.com/Agriculture/68699298-Success-stories-on-information-and-communication-technologies-for-agriculture-and-rural-development.html (accessed Jan 30, 2018).
36. Tank, K. 2018. http://www.sciensopedia.com/tecnology-in-agro-field-boosts-development/ (accessed Feb 3, 2018).
37. TataConsultancyServices. 2017. https://campuscommune.tcs.com/en-in/intro/viewblog/tcs-features-in-fortune-50-list-of-companies (accessed Jan 30, 2018).
38. The Food and Agriculture Organization (FAO). 2000. http://www.fao.org/docrep/004/y0937e/y0937e03.htm (accessed Feb 12, 2018).

39. TNAU Agritech Portal. 2015. http://agritech.tnau.ac.in/pdf/ICT20success20stories.pdf (accessed Feb 12, 2018).
40. United Nations Department of Economic and Social Affairs Division for Social Policy and Development. 2013. www.un.org/esa/socdev/egms/docs/2013/ict/CraigLehmann.pdf (accessed Feb 12, 2018).
41. United Nations High Commissioner for Refugees. 2018. http://www.unhcr.org/ 47f0a6db2.pdf (accessed Feb 12, 2018).
42. United Nations Organization. 2009. www.fao.org/wairdocs/am087e/am087e.pdf (accessed Feb 12, 2018).
43. Vachani, S.; Smith, N. C. Socially Responsible Distribution: Distribution Strategies for Reaching the Bottom of the Pyramid. *Calif. Manage. Rev.* **2008,** *50* (2), 52–84.
44. Vodafone Group. 2015. https://www.vodafone.com/content/dam/vodafoneimages/sustainability/downloads/54909_Vodafone_Connected_Farmers_Final.pdf (accessed Jan 20, 2018).

CHAPTER 3

Potential of Mobile Phones and Social Media Platforms in the Agricultural Scenario

KELVIN JOSEPH BWALYA

ABSTRACT

The development of mobile technology and ultimately increased participation of individuals in social media platforms have translated into increased individual's inclusion in many different socioeconomic undertakings. In the African context, agriculture has shown trends of benefiting in terms of farmer's access to agriculture markets, financial markets, cardinal agriculture innovations, and context-aware information. Marginal smallholder farmers in geographically disadvantaged locations, who are normally secluded from meaningful socioeconomic value chains, are integrated into information highways owing to their wider adoption of mobile technologies and social media platforms. Since many of the countries in the Sub-Saharan Africa depend, to a large extent, on agriculture, it is important that the different actors and farmers should be given an opportunity to access different opportunities and information in the agriculture sector so that they can make informed decisions at any given time. The proliferation of adoption and usage of mobile technologies and social media platforms will culminate into more access to information and opportunities by the smallholder farmers.

3.1 INTRODUCTION

From time immemorial, agriculture has been the mainstay of many economies in Africa and the wider developing world contexts. The importance of the agricultural sector cannot be overemphasized as it is a source of livelihood for over 75% of rural people in rural Africa. Unfortunately, the same rural areas are mostly hit with poverty and underdevelopment (lack of requisite socioeconomic infrastructure such as impassable roads, etc.). Many of the African farmers are lagging behind to bring to the market high-quality agricultural produce due to limited access to innovations in agriculture technologies, seeds, farm management styles, etc., and lack of access to scientific innovations at different stages of the agricultural value chains. Generally, limited or lack of appropriate agriculture technologies and market information account are one of the key reasons for limited agriculture development in Africa.[1] This is unfortunate as most of the Sub-Saharan Africa (SSA) is considered to have fertile soils, right kind of rainfall, good weather for animals, etc. The aforementioned challenges have contributed toward continued use of old and outdated farming styles by smallholder farmers, who were negatively impacted for contemporary farming styles.

In order to avert these adverse effects, many stakeholders in the agriculture sector have come up with strategic initiatives to ensure that there is access to agriculture information and innovations at all levels of the agriculture hierarchy. These initiatives are being complemented with the speedy development of mobile technologies and social media applications in many socioeconomic contexts of Africa.

In Africa, the mobile penetration rate in 2007 was 41%.[1] Currently, this has almost doubled. Africa has shown higher mobile penetration rate with over 960 million mobile subscribers. This represents roughly about 80% of the total continent's population. There are total of 216 million Internet users in Africa representing 18% of Internet penetration. Mobile phones are indispensable tools in disseminating and sharing agricultural information pertaining to different aspects of agriculture such as seeds, strategies and practices, and markets. In the African context, there is an increased use of social media in the dissemination of indigenous agriculture knowledge in sharing of practices with both local and international audience on the traditional agriculture practices.[23]

Because information is one of the key ingredients for competitive agriculture, the need for requisite information management models in agriculture cannot be overemphasized.[28] A study investigating the transactional costs in agriculture value chains in Sri Lanka showed that 70% of the transactional costs was attributed to information costs.[4] The cost that comes with the neglect of information as a key resource for agriculture cannot be imagined. In the information-intensive setup of contextual environments, it can be affirmed that the emphasis and cost of information have tremendously increased. In order to mitigate the costs of information management in agriculture value chains, many organizations are transforming their organization's data management roles to include cloud and fog computing.[6,22]

This chapter explores the current and future scenarios of usage of mobile phones in different agriculture setups in Africa for sharing of information at the center of agriculture innovation. The specific objectives are[5]:

- To examine actual levels of adoption and usage of mobile phones and social media in agriculture;
- To understand issues at the center of mobile phones and social media usage in agriculture; and
- To understand the context in which mobile phones and social media can be appropriately utilized in the agriculture value chains in SSA.[5]

This chapter is based on a rigorous literature review of academic journal articles, conference papers, and public technical reports and working papers principally from international organizations on agricultural strategies. Databases involved in the search were: ABI-Info (ProQuest), Science Direct, Emerald Fulltext, EBSCO Business Source Premier, and Google Scholar. The narrative review is important because there are currently limited studies on adoption and usage of mobile phones and social media platforms focusing on agriculture in Africa. Synthesizing the available information in this area and giving informed insights is very important to explore the possible scenarios for the usage of mobile phones and social media in sharing of innovations and practices surround contextually relevant agriculture practices.[2,9]

3.2 INFORMATION SHARING IN AGRICULTURE CONTEXT

Information sharing among smallholders using emerging social media applications enables them to harness the different opportunities that pop-up in agricultural establishments. Information sharing is important in contemporary agriculture setups especially given the precedence of ever-changing climate patterns throughout the world. The changing climate patterns bring about degradation or declining trends in soil fertilities and calls for scientific information sharing Integrated Soil Fertility Management (ISFM) information and knowledge access and sharing. Information sharing is very important in different agriculture environments so as to overcome the different challenges faced in contemporary agriculture environments.

Without contemporary and effective information sharing, smallholder farmers miss out on technology innovations to help them maintain competence in the agriculture continuum. A more accessible (almost globally) and convertible social media platform is desired for farmers to participate in the agriculture information value chains.[15] Further, lack of appropriate information dissemination and sharing platforms culminates into a gap between the knowledge held by the farmers and the researchers and this has a direct impact on the level of innovation in a given context.[21] There are key factors that need to be considered when designing any information sharing interventions. Figure 3.1 shows the key factors in effective information sharing in agriculture contexts. This figure is conceptualized upon the literature review explored.

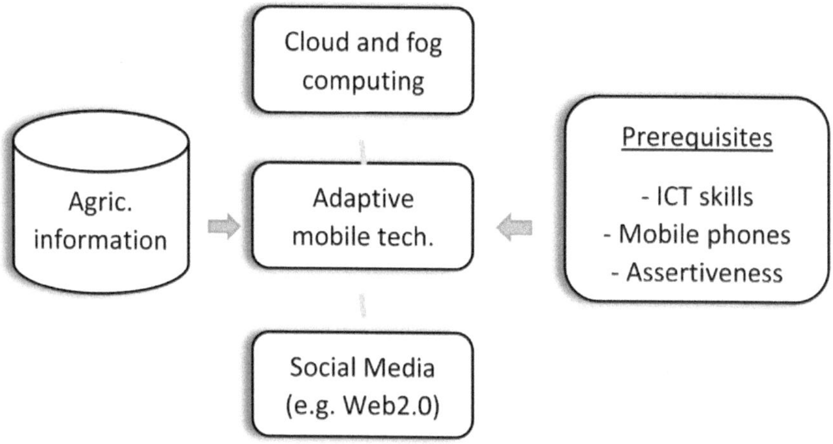

FIGURE 3.1 Key steps for effective agriculture sharing.

The constructs shown in Figure 3.1 need to be considered as drivers in the formulation of policy to guide information sharing in agriculture establishments. The conceptualized information sharing ecosystem is a three-level model:

- The first level is the agriculture information content and innovations available in the social network or in different organizations that needs to be shared for wider utilization.
- The second level shows the different information-sharing enablers, platforms, and systems, which are key ingredients for sustainable agriculture information sharing. Some of these enablers include adaptive mobile technology gadgets and platforms, which can easily be accessed by different types of farmers. Further, the emergency of cloud and fog computing platforms enables network-based agriculture information to be easily accessed ubiquitously. Social media platforms allow the different stakeholders to share information easily in mainly informal ways.
- The third level focuses on the characteristics of the individual farmers. A farmer needs to be endowed with appropriate ICT skills, have a mobile device which is Internet-enabled, and be assertive to sharing and receiving different types of agriculture information to be employed in the different agriculture initiatives.

3.3 AGRICULTURE INFORMATION AND INNOVATION

The sharing of different information among different players in the agriculture value continuum allows innovation to be realized in different contextual settings. Apart from economic and personal considerations, the desire to innovate is also motivated by social interactions with the change agents in the environment in which they operate. Some change agents in the agricultural value chains include: agro-dealers, agriculture input suppliers, researchers, and farmer's associations.[12] Studies targeting natural resource problems have been solved by conceptualizing innovative solutions based on the farmer's social network structure. Therefore, it can be said that the more connected a network of farmers is, the more likely it is to share innovations and knowledge at the forefront of global knowledge frontiers and appropriately react to any contextual challenges that may be faced. In traditional information sharing environments, sharing ideas is very difficult. Contemporary information networks have been boosted by wider penetration of social media networks as an easy and pervasive information management environment. The strength and effectiveness of a network is a spatial-temporal function of emotional intensity, intimacy, and reciprocity.[11]

3.4 NETWORKING AND AGRICULTURE

The conceptualization of a network entails that individual parts of the network (nodes) are able to share information about their functional state and their capacity to contribute to the overall goal of the network.[8] In the agricultural environment, network nodes (individuals) share information about their experiences, knowledge obtained, or occurrences in the agricultural value chains. Social networks can be identified as operational networks, personal networks, and strategic networks. One of the most common networks proving to be more effective is the social network. Social networks are advantageous in that they also enable sharing of private information between individuals, enable individuals to benefit from knowledge generated out of the availability of diverse skill sets—this is important as contemporary competitiveness is garnered from access to contemporary knowledge found in modern organizations which are highly inter-and transdisciplinary,[6] and are power gained.

In agricultural setups, the different types of networks exist in the form of knowledge and dissemination networks.[8] These networks are cardinal in the sharing of knowledge. The climate adaptation (e.g., how to avert the effects of greenhouse gas emissions, global warming, sea level increase, etc.), emergence of eco-friendly energy solutions and innovations (e.g., renewable energy), biotechnology, etc.,[4] have transformational adaptation to changing climate conditions that are likely to be achieved if farmers belong to effective social networks that may be small, large, dense, and wide-spread information networks.

3.5 BENEFITS OF SOCIAL MEDIA

Social media may encompass different things in different contexts. In general, social media encompasses a combination of user-generated information content, various communities, and Web2.0 technologies, defined on the action of the users.[7] The gamut of social media should include: networking sites, RSS feeds, blogging sites, wikis, podcasting, and 3D virtual environments.

There are many benefits attributed to social media networking:

- Social media permeates through the different agriculture socio-economic boundaries and allows information to navigate and flow through diverse farming blocks.
- Contemporary social media can be used in ensuring that there is requisite information in the different agricultural eco-systems.
- Farmers' engagement of social media as a communication platform depends on the communication dynamics of information dissemination that exists between the communication parties and the characteristics and ability of the individual farmer to actively participate in the agriculture information highways.
- Effective information and knowledge sharing need to be pursued in any given context to the effect that there are tangible communication channels between information designers and implementers.[3]
- Inclusive and participatory information and knowledge management sharing by all actors is desired at all stages of the agriculture continuum. Therefore, the need to circulate, communicate, and aggravate new innovation styles and systems in agricultural research cannot be overemphasized.

The different uses of social media in the realm of farmers' research have been articulated on different aspects of agriculture and also as a tool for agricultural marketing. Generally, farmers' use of social media translates into reducing the time and cost for obtaining relevant and useful agricultural information. Some uses of common social media tools are:

- *Blogs*—individual online presence, where a person has a Web diary where he or she shares information detailing experiences and personal opinions on a given topic. May have detailed biographical information, photos, and links to other websites of common interest (e.g., RSS feeds). Common platforms for designing and creating web blogs are www.wordpress.org and www.blogger.com
- *Facebook*—although rocked with intermittent data leak scandals, recent being the Cambridge Analytica scandal, yet Facebook remains the most common and popular social media platform for profiles and pages. Recently, a lot of marketing activities and campaigns have been taking place on Facebook.

- *Flickr* (www.flickr.com) is a platform mainly for managing and sharing photos either in the private or public mode. The site gives an opportunity to rate or rank photos displayed thereon.
- *LinkedIn* is basically a business networking site, which gives a platform for professionals to prove a snapshot of their skills and interests. Can be used to make professional networks, find advertised jobs, business opportunities, people, etc.
- *Twitter* (www.twitter.com) is a maximum 140 character micro-blogging platform to broadcast short and straight-to-the-point messages to followers. Users have the opportunity to choose, whom to follow, what messages from other users to re-tweet, etc.
- *WhatsApp* comes in handy for person-to-person communication and is a preferred medium for related groups.
- *YouTube* videos are for information accessed through different applications. Social events or long music videos can be captured and uploaded on YouTube.

Social media platforms can be used to connect farmers with fellow farmers in a bid to exchange information, and agricultural experts for their engagement in the innovation value chains.

3.6 POTENTIAL OF MOBILE PHONES IN AGRICULTURE VALUE CHAINS

The potential of mobile phones in agriculture value chains cannot be overemphasized. Mobile phones are used to support activities for marketing, quality assurance of produce, and market access.[10] Farmers are able to connect with their potential markets and announce their produce before it is actually taken to the market and, in some instances, before it is actually grown. This enables the assessing of produce's potential markets before they are actually taken to the market as a result of direct interaction with the suppliers, traders, end-user markets, extension services, etc. Mobile phones will enable farmers' engagement in different contextual supply chains for new forms of organization, community engagement, learning, and interaction.

Because of recent advances in mobile phone technology platforms, they are now used to provide services, which have culminated into farmers' increased access to market information (different dynamics that define the market structure and politics), and therefore improved coordination of input/output resources utilized in transport and logistics, finance, and production techniques.[1,10]

3.7 SMART AND E-AGRICULTURE

The role of social media in innovation sharing especially given the emergence of smart and e-Agriculture even in developing world contexts cannot be overemphasized. As said before, information sharing is becoming very critical in contemporary agriculture environments, where many players have continuously been showing a desire to jump onto the bandwagon. The changing dimensions in the agricultural sector throughout the world are being exacerbated by ever-changing climate conditions partly due to global warming. Consequently, there have been many challenges that have surfaced threatening the wider global food security. Therefore, farmers are implored to employ smart-agriculture techniques and innovative solutions to overcome the different eminent challenges such as greenhouse gas emissions, farmers' vulnerability to climate change, etc.

If information sharing is not effective in any farming environment, farmers in that farming block will miss out on the latest trends and innovations in farming. A conceptual framework, climate-smart agriculture (CSA), has emerged in recent years to collate efforts to abate many challenges faced in agriculture. CSA is being adopted even in the developing world contexts.[17] For example, in the East African region, farmers are continuously adopting requisite CSA practices to mitigate the vulnerabilities to changing climate conditions.[20]

In case of East Africa, and a major parts of Africa, smallholder farmers do not have the financial muscle to procure expensive irrigation equipment or air-conditioning equipment (in case of livestock farmers) in order to avert the effects of climate change and drought. In many parts of arid Africa, many livestock have died due to extreme (high) temperatures, therefore, livestock farmers may opt for air-conditioning equipment in order to keep the temperatures down in animal shelters. The type of air conditioners desired cannot be known if there is no effective information sharing among farmers. Social networks may be used to share word on technology innovations and best practices of CSA given the context.

3.8 MOBILE NETWORKS IN AGRICULTURE: AFRICA

Due to higher adoption rates, mobile phones stand as one of the potential tools for revitalization of communication and access a wide array of information. Realizing the potential of mobile technologies, many farmers in SSA are positioning themselves toward adopting and using contemporary mobile technologies and networks. For example, as of 2012, 63% of Nigerian farmers had adopted Internet-enabled mobile phones as a reliable communication and information access platform.[11] The trend is the same across Africa with most countries having higher mobile phone penetration rates. The increased adoption of cheaper 3G mobile phones in Africa has culminated into improving farmer-to-farmer communication, where agriculture information can be transmitted by voice and video.[27] Currently, over 60% of Africa's economy is dependent on agriculture and, therefore, there is need to share agriculture information as much as possible to have competent agriculture systems.

As the mainstay of the economy employing around 70% of the national workforce, agriculture contributes over 40% of Nigeria's gross domestic product (GDP).[26] Because it is an important part of the socioeconomic infrastructure, agriculture establishments have largely adopted social media as a platform for information interchange between researchers and farmers so that the designers and farmers have access to latest innovations that could ultimately improve their farming outputs. Therefore, there is need for current and context-specific information in order to drive competitiveness in the different agriculture contexts. The higher adoption of different forms of ICTs enables requisite dissemination and flow of information influencing different decisions and strategies made. The ICT platform, which has largely been adopted in the context of Nigeria, is the social media that is perceived as an excellent platform for tagging, blogging, networking, discussion, etc.

The degree of penetration of mobile technologies and social media has gone to all corners of Africa in such a manner that over 80% of the people are using one form or the other of mobile phones. The increased penetration of mobile technologies is further encouraged by the adaptive design models, which take into consideration different types of persons, for example, a disabled person can interactively choose the way in which information needs to be presented to him or her. Contemporary social media designs are presented to individuals based on their characteristics.

Adaptive models make it possible for global inclusion of individuals into the agriculture information value chains. By using mobile devices, smallholder farmers in Africa are now able to know beforehand the price of commodities on the market instantly unlike a long time ago when they had to travel long distances just to go and understand the price structure of their farm produce.[13] Further, farmers are now able to appropriately respond in time to turbulent agriculture market dynamics.

With over 30 million mobile phone subscribers in South Africa, innovative solutions targeting the mobile phone information space can reach a lot of people.[22] With the active participation of Independent Communications Authority of South Africa (ICASA) in regulating the mobile phone market, the cost of effective use of mobile phones is significantly down and is affordable to majority of people.

As aforementioned, many countries in Africa have agriculture as one of the major contributors of GDP. For example, in Rwanda, agriculture is one of the key economic activities providing employment to over 70% of the workforce. The steps needed in the design and implementation of a "climate smart agriculture and sustainable intensification" have been articulated in Rwanda.[25] The design is hinged to get better data management practices among the smallholder farmers and different stakeholders with regards to integrated water management, soil fertility levels, quality adaptable seeds (cereals or legume crops), capacity building, knowledge management, etc.[11]

In order to develop the agriculture sector, the Southern Africa Development Community (SADC) came up with the SADC Regional Indicative Strategic Development Plan (RISDP) to speed up the agriculture development in a bid to effectively contribute to economic development. The RISDP has translated into increased agriculture production and emergence of smart farmers across the region. Upon realizing many benefits of mobile technologies in the agriculture and other sectors, many multi-stakeholder projects are being implemented around Africa. Selected promising country-level or multi-country-level projects are as follows:

- *Chipata Women's Mobile SMS* project proposed by OneWorld Africa targeting rural women in Chipata district of Zambia: Focuses on information exchange on agriculture markets and strategies for public reduction. This projects aims to achieve sustainability of smallholder farmers by providing them with relevant, current, and usable information on contemporary agriculture practices;

- *Dokoza* proposed by the State Information Technology Agency (SITA) and other institutions: It targets the health sector focusing on data exchange of real-time health data and using the mobile phones as the key enabler;
- *Xam Marsé SMS market information service* in Senegal aims to provide farmers with agriculture information (prices, demand for fresh products, etc.) for sustainable development;
- In the context of *Mali and Burkina Faso,* mobile phones are being actively used by extension officers as media for information and knowledge transfer overcoming the frequently experienced distance barriers.[27] Extension officers are able to send video information through 3G mobile phones thereby easily putting across their messages to the farmers and vice versa. Further, the use of visual information appropriately extends the experience of farmer-to-farmer information.
- In *Uganda,* the lifelong learning mobile App provides information to farmers in rural areas on emerging farming techniques and new knowledge on general trends in farming given the context.[14] These different initiatives have gone a long way in bridging the digital divide allowing more people to use mobile phones and social networks to ubiquitously share information;
- *Maluleke project in South Africa* aims to provide market price information of various farm products to smallholder farmers in Limpopo province;
- *South African Depression and Anxiety Group* (SADAG) targets counselling on teen suicide;
- *Women of Uganda Network* (WOUGNET) aims to do electronic delivery of agriculture information especially market related information to the rural communities in Uganda[16];
- the *Village Phone Initiative* in Uganda, Rwanda, and Cameroon is led by MTN to reduce the cost of telecommunications to the rural poor people so that they can participate in the different information management value chains;

The effect of social networks in increasing agricultural innovations and social networking has been investigated in agriculture environments in Ethiopia.[18] The social network configuration and structure is the precursor for agriculture information in any context. The more the connection is meshed, the more the newer information can easily permeate the network and ultimately reach the farmers for incorporation into their agriculture practices.

The UK nonprofit organization Mercy Corps is piloting a mobile App, Agri-Fin, in Uganda and Zimbabwe to help farmers obtain desired agricultural information, such as which seeds to plant, buy crop insurance, pests, fertilizers, market information, etc. The App specifically helps farmers to access pervasive information without having to walk long distances to access the services of agriculture information services. Agri-Fin helps bring together the farmers and available financial services directly supporting agricultural information. In conclusion, Agri-Fin provides the needed information space bringing together farmers, banks, agriculture research institutes, farmer organizations, etc., creating the information comprehensive environment allowing farmers to conveniently and effectively engage in farming activities. The overall benefit of this may be increased agricultural productivity.

The government and other stakeholders in South Africa have targeted strategies to promote the inclusion of smallholder farmers into the mainstream agriculture value chains.[19] This is strategically important because farming accounts for about 3% of the GDP and employs over 4.7 million workers. With the rising unemployment in South Africa (currently at 26.7%), putting more information in the public domain is one of the most effective ways to encourage many people to engage in agriculture as an economic activity. The more information is in the public domain, the more people will understand what it takes to engage in farming. Important information may include: source of start-up capital for procuring farm equipment, assessment of markets, labor dimensions in farming setups, farming practices, farming communities, etc. Recognizing the value of agriculture information especially among the smallholder farmers, the Government of South Africa has invested substantial amounts of money in ensuring that agriculture extension officers are in every ward for easy access of agriculture information. In the contemporary environment, social media, as in many other contextual settings, presents itself as one of the most effective vehicles to promote the proliferation of access to information on public domains.

Another critical intervention when dealing with smallholder farmers is that they need to be involved in any interventions ear marked for them. It is a common fact that innovations made from expensive R&D efforts made without the consultation of smallholder farmers may be unsuitable to their contemporary needs given the context in which they operate. The exclusion of smallholders in the design of interventions for smallholders ultimately leads to failure or low adoption of such innovative solutions.[19] Locally grounded technology innovations hinged upon absolute involvement of smallholder farmers stand a chance for meaningful adoption. In order to encourage absolute adoption, it is important to bring together the designers and smallholder farmers in one space to exchange ideas and needs. This space should not necessarily be a defined temporal-spatial space but can be a virtual space brought about by social connections that are in turn brought about by ICTs. Social media is a good candidate in this aspect as it allows farmers located even in the remotest parts of Africa given its ubiquitous characteristic to communicate with the desired individual anytime and anywhere.

Most of African countries have done well in providing information to farmers using state-sponsored display units where agriculture marketing information and other types of information pertaining to agriculture are displayed. The only downside of display panels is that most of them are usually mounted in locations where farmers in remote places cannot access them and therefore are left out from the information highway. Mobile phones overcome the special distance among different communities and farmers are able to access recent innovations, market information, general agriculture information, etc., in the comfort of their homes. By doing so, mobile technologies unlock unimagined opportunities to the benefit of the farmers.

3.9 FUTURE OF SOCIAL NETWORKS

Not taking advantage of many of the emerging technologies such as new forms of social media, many SSA countries have continued to rely on extension officers as the fountains of agriculture information. But due to varying cultural practices, many of the female smallholder farmers do not appropriately benefit from agriculture extension because many of the extension officers are males. Other smallholder farmers are completely cut off from agriculture information value chains because they farm in locations with poor road networks and therefore are not accessed by agriculture extension workers.

Social networks of the future will involve the rewarding of content contributors on the social space and ecosystem. The ultimate benefit of such a scenario would be increased sharing of agriculture information in the social networks with the anticipation that there will be rewarded for sharing knowledge which will eventually culminate into competitive agriculture. The Mithril (MITH) token in the Asian countries of China (Hong Kong and Taiwan) can be used to obtain network value which can further be monetized as a way to reward content creators. MITH tokens can be used as payment method in retail shops belonging to Mithril Merchant Networks partners. There is no doubt that in future there will so many adaptive applications that can be used in different contexts.

With advances in virtual and augmented reality, it is anticipated our lives will be digitized as much as possible. In the farming realm, social media will be embedded with augmented reality in such a way that mobile phones will be used to interactively locate potential markets of farmers' produce within a certain location. There will be more agriculture information on social networks that situations and solutions will be a click away.

The crowdsourcing will go a long way in ensuring that many information sources are integrated to come up with one set of comprehensive information. Crowdsourcing will bring convenience to social networkers in that many everyday applications such as buying cars, air tickets, etc., will be sorely accomplished online. The farming community will source for funding of their agriculture initiatives using adaptive online applications.

With the continued development of emerging technologies such as cloud and fog computing, and smart cities, ambient technologies will enable enhanced communications blended into surroundings ultimately shaping the context in which information is shared. In South Korea, the network nodes in Taejon enable network users to virtually communicate with their surroundings pointing them to nearest transport stations and telling them how long they will have to wait before the next bus or train arrives. Such possibilities will go a long way when applied in agriculture setups.

Network-based research is slowly becoming very popular. With the ever-development of Big Data and predictive analytics, researchers will be able to instantaneously analyze information on social networks as it is created coming up with up-to-date trends in their area of enquiry. In the agriculture realm, researchers will be able to make available current information on trends and different aspects of agriculture development. By putting this information online, farmers will be able to access current information and thereby improve their agriculture applications.

In order to encourage participation in different social networks, it is anticipated that future social networks will have more games introduced into the different social ecosystems to increase user engagement. The possibility of a highly interactive social media space is possible given the ever-increasing capabilities in broadband connections and variety of mobile technology platforms. The portability of mobile devices and increase in cloud and edge computing will open up a lot of opportunities in social media platforms which will ultimately benefit the smallholder farmers in remote locations.

There are over 1.5 billion social networking users in the world,[10] thus accentuating the potential of reachability of social media platforms in agriculture information. Social media platforms increase the value of social capital by bringing together industry, the farmer, and consumer so that there is requisite sharing of information. Social networking enables farmer-to-farmer networking with a bid to reducing social isolation of farmers thereby encouraging social cohesion so that there is exchange of ideas and farming experiences. Social media gives farmers a platform to lobby other industry players and come up with collective voice, which culminate into collective actions. Further, social media promotes consumer engagement (farmer or industry and consumers) so that there is clear understanding of interest.

3.10 FUTURE PROSPECTS

Although mobile technologies have proved to be indispensable tools for effective communication in the contemporary world, yet they also have side effects. For example, excessive use of mobile phones as means for agriculture information interchange may culminate into seclusion of a sizeable number of farmers due to limited or lack of general or ICT literacy skills. In the future, there will be more adaptive social media platforms. which will increase the general usability of platforms to the benefit of smallholder farmers. Further, the emergence of Web2.0 and semantic Web variants will change the interaction models of individuals on social networking sites. Already, Web2.0 is being used in different agriculture domains. For example, the use of Web2.0 as an information-sharing platform for assessing agriculture trends and shaping decisions has been investigated in the different agriculture value chains.[7]

Future studies must investigate the different social dimensions that influence the adoption and use of social media platforms in the African context. Supplying the right mobile technologies and platforms is one and ensuring that farmers adopt and use those platforms is another. A study in Malawi[4] has shown that exchanging information through social media platforms from one farmer to another is not automatically given and may require greater extra effort in order to make it happen. Readers are encouraged to consult the book *Data Analytics and Big Data* by Soraya Sedkaoui.[1]

3.11 SUMMARY

This chapter explores the literature focusing on famous databases such as Scopus, Emerald, etc., to understand the contemporary trends in the usage of mobile technology applications and social media platforms among smallholder farmers in SSA. This chapter has showcased the importance of mobile technologies and social media platforms in facilitating ubiquitous sharing of agriculture information among farmers and different industry stakeholders. Social media is arguably one of the best platforms that can be used in the sharing of information pertaining agriculture by smallholder farmers.[24] It is clear from the study that there are different efforts being pursued in different contextual settings to advance ubiquitous access to current and useful agriculture information.

[1] Sedkaoui, S. *Data Analytics and Big Data*, 1st ed, Format Kindle; Wiley-ISTE: New York, 2018; p 220.

KEYWORDS

- **Africa**
- **agriculture**
- **communication technology**
- **mobile phone**
- **mobile technology**
- **social media**
- **Sub-Saharan Africa**

REFERENCES

1. Asenso-Okyere, K.; Mekonnen, D. A. The Importance of ICTs in the Provision of Information for Improving Agricultural Productivity and Rural Incomes in Africa. http://www.africa.undp.org/content/rba/en/home/library/working-papers/icts-provision-information.html (accessed Jun 22, 2018).
2. Balkrishna, B. B.; Deshmukh, A. A. Study on Role of Social Media in Agriculture Marketing and Its Scope. *GJMBR*. **2017,** *17* (1), 1–5.
3. Ballantyne, P. Agricultural Information and Knowledge Sharing: Promising Opportunities for Agricultural Information Specialists. *Agric. Info. Worldw.* **2010,** *3*, 1–4.
4. BenYishay, A.; Mobarak, M. Communicating with Farmers Through Social Networks. *Afri. J. Hospital. Tour. Leisure* **2017,** *6* (1) (e-article).
5. Chilonda, P.; Minde, I. Poverty, Food Security and Agricultural Trends in Southern Africa. Working Report 1; Regional Strategic Analysis and Knowledge Support System for Southern Africa (ReSAKSS-SA), Silverton 0127 Pretoria, South Africa; 2007; p 67.
6. Chisenga, J.; Kedemi, R.; Sam, J. The Use of Social Media in Agricultural Research Workflows in Ghana and Kenya. *Agric. Info. Worldw.* **2014,** *6*, 1–7.
7. Chisita, C. T. Knotting and Networking Agricultural Information Services Through Web 2.0 to Create an Informed Farming Community: A Case of Zimbabwe. 2012. World Library and Information Congress, 78th IFLA General Conference and Assembly, Helsinki. https://www.ifla.org/past-wlic/2012/205-chisita-en.pdf (accessed Jun 22, 2018).
8. Cunningham, R.; Jacobs, B.; Measham, T.; Harman, M. P.; Cvitanovic, C. *Social Network Analysis. A Primer on Engaging Communities on Climate Adaptation in New South Wales, Australia;* UTS: ISF, Australia, 2017; p 69.
9. Gaillard, J. C. Vulnerability, Capacity and Resilience: Perspectives for Climate and Development Policy. *J. Int. Devel.* **2010,** *22*, 218–232.
10. Duncombe, R. *Mobile Phones for Agricultural and Rural Development: A Literature Review and Future Research Directions*. Report No 1; Centre for Development Informatics Institute for Development Policy and Management, SED University of Manchester, Arthur Lewis Building, Manchester; **2012;** p 38; ISBN: 978-1-905469-31-4.
11. Granovetter, M. S. The Strength of Weak Ties. *Am. J. Sociol.* **1973,** *110*, 1360–1380.
12. Hartwich, F. S.; Scheidegger, U. *Fostering Innovation Networks: The Missing Piece in Rural Development*. Rural Development News No 1; Swiss College of Agriculture, Switzerland, 2010; p 61.
13. Karume, S. M. Transforming Governance Through Mobile Technology in Developing Nations: Case of Kenya. *Comp. Infor. Sci.* **2016,** *9* (4), 8.
14. Lwasa, S.; Asingwire, N.; Okello, J. J.; Kiwanuka, J. Awareness of ICT-based Projects and Intensity of Use of Mobile Phones among Small Holder Farmers in Uganda: The Case of Mayuge and Apac Districts. *Int. J. ICT Res. Devel.*, **2010,** *2* (2), 26–38.
15. Mashavave, T.; Mapfumo, P.; Mtambanengwe, F.; Gwandu,T.; Siziba, S. Interaction Patterns Determining Improved Information and Knowledge Sharing Among Smallholder Farmers. *Afr. J. Agric. Res. Eco.* **2016,** *8* (1), 1–12.
16. Masuki, K. F.; Tukahirwa, J.; Kamugisha, R.; Mowo, J.; Tanui, J.; Mogoi, J.; Adera E. O. Mobile Phones in Agricultural Information Delivery for Rural Development

in Eastern Africa: Lessons from Western Uganda. Unpublished Report, 2016; p 11. http://www.worldagroforestry.org/downloads/Publications/PDFS/MM10320.pdf (accessed Jun 22, 2018).
17. Meinzen-Dick, R.; Bernier, Q.; Haglund, E. The Six "Ins" of Climate-smart Agriculture: Inclusive Institutions for Information, Innovation, Investment, and Insurance. CAPRi Working Paper No. 114; International Food Policy Research Institute, Washington, D.C.; **2013;** pages 11; Online: http://dx.doi.org/10.2499/CAPRiWP114 (accessed Jun 22, 2018).
18. Mekonnen, A. D.; Nicolas. G.; Anna. J. A. Social Networks, Agricultural Innovations, and Farm Productivity in Ethiopia. Working Paper Series N° 235; African Development Bank, Abidjan, Côte d'Ivoire, 2016; p 34.
19. Mudhara, M. Agrarian Transformation in Smallholder Agriculture in South Africa: A Diagnosis of Bottlenecks and Public Policy Options. Paper presented at 2010 Agriculture conference "Overcoming Inequality and Structural Poverty in South Africa: Toward Inclusive Growth and Development," Johannesburg; Sept 20–22, 2010; p 9.
20. Nyasimi, M.; Kimeli, P.; Sayula, G.; Radeny, M.; Kinyangi, J.; Mungai, C. Adoption and Dissemination Pathways for Climate-smart Agriculture Technologies and Practices for Climate-Resilient Livelihoods in Lushoto, Northeast Tanzania. *Climate* **2017,** *5* (3), 63. https://doi.org/10.3390/cli5030063.
21. Odendo, M.; Ojiem, J.; Bationo, A.; Mudeheri, M. On-farm Evaluation and Scaling-up of Soil Fertility Management Technologies in Western Kenya. *Nutr. Cycl. Agroecosys.* **2016,** *76*, 369–381.
22. Okello, J. J.; Ofwona-Adera, E.; Mbatia, O. L. E.; Okello, R. M. Using ICT to Integrate Smallholder Farmers into Agricultural Value Chains: The Case of Drumnet Project in Kenya. *Inter. J. ICT Res. Devel. Afr.* **2010,** *1* (1), 23–37.
23. Owiny, S. A.; Mehta, K.; Maretzki, A. N. The Use of Social Media Technologies to Create, Preserve, and Disseminate Indigenous Knowledge and Skills to Communities in East. *Int. J. Comm.* **2014,** *8*, 234–247.
24. Poulton, C.; Doward, A.; Kydd, J. The Future of Small Farms: New Directions for Services, Institutions and Intermediation. *World Develop.* **2010,** *38* (10), 1413–1428.
25. Prasad, P. V. V.; Hijmans, R. J.; Pierzynski, G. M.; Middendorf, J. B. Climate Smart Agriculture and Sustainable Intensification: Assessment and Priority Setting for Rwanda. Feed the Future Sustainable Intensification Innovation Lab Final Report; Kansas State University, Kansas, USA, January 2016. http://hdl.handle.net/10568/87987 (accessed Jun 22, 2018).
26. Sokoya, A. B.; Onifade, F. N.; Alabi, A. O. *Establishing Connections and Networking: The Role of Social Media in Agricultural Research in Nigeria.* World Library and Information Congress, 78th IFLA General Conference and Assembly, Helsinki, 2012; p 98.
27. Sousa, F.; Nicolay, G.; Home, R. Information Technologies as a Tool for Agricultural Extension and Farmer-to-farmer Exchange: Mobile-phone Video Use in Mali and Burkina Faso. *IJEDICT* **2016,** *12* (3), 19–36.
28. Stigler, G. The Economics of Information. *J. Pol. Eco.* **1961,** *69* (3), 213–225.

CHAPTER 4

Role of Social Media in the Socioeconomic Development of Africa

KEHBUMA LANGMIA

ABSTRACT

Social-media platforms have now become the new communicative arena for humanity across all spheres of life. This new electronic mode of communication has introduced virtuality as the new frontiers for debating socioeconomic issues affecting mankind. This study critically assesses existing literature on the subject of social media role in the continent of Africa, and how they have enhanced developmental needs or the lack thereof. Results show that a new platform of communication in fact facilitates socioeconomic development in Africa. However, there are still issues like digital divide that seems to impede progress.

4.1 INTRODUCTION

Alfred Lord Tennyson (1809–1892) must have had a vision when he said, "*The old order changeth yielding place to the new,*" in his poem "Morte D' Arthur." He was far ahead of the renowned communication Canadian scholar, Marshall McLuhan who, in the 1960s predicted that electronic communication was going to surpass in-person communication.[14] They were, unbeknownst to themselves, foreshadowing the "sudden" arrival of the Internet. The Internet has ushered another innovative interactive-communication tool called the social media. Social media contain social networking sites (SNS). SNS are platforms for human-to-human communication like Facebook, Twitter, Instagram, LinkedIn, Skype, Tumblr, Pinterest, WhatsApp, Viber, Snapchat, etc. Relying on the Internet for connectivity around the world, the social media are attempting to carve a new communicative route for themselves. As the new electronic communicative platforms that have galvanized individuals from all backgrounds across the globe, social media are redefining human communication.[7,10,11,15]

This century is echoing a much anticipated communicative refrain: You are in or you are out! No mercy! Africa, the continent that has seen pejorative adjectives heaped on her in the wake of the forceful 1884 Berlin Conference colonial conquest by the West, is waking up to a dawn—the dawn of striving to "belong" to the "G-20s" global forces. As a result of the fact that Africa was behind in the industrial revolution, she wants to be with the rest of the world in the Internet revolution. This is exactly why the acquisition rate of mobile phones in the continent is geometrically increasing very fast.

According to the International Telecommunication Union (ITU) May 2014 Press Release (see www.itu.int), Africa and other developing countries have a mobile phone penetration rate of about 80%. This is a significant development given that by 2004, the rate was lower compared to the rest of the world. Nowadays Africa is bypassing the rest of the world in their penetration rate. It is with these mobile phones that users in the continent are able to perform electronically related services like e-learning and e-banking activities across state boundaries. This is unprecedented, and as a result, other information communication technology (ICTs)-related activities are being performed socially, culturally, and economically; thereby, uplifting standards and lowering costs of living for ordinary folks across the continent.

This chapter demonstrates the power and influence that social media have on socioeconomic development of the continent in the last 10 years. The research questions that will be used to seek answers to the issues of social media and socioeconomic development of Africa are: How has related literature on social media address socioeconomic imperatives of Africa? To get appropriate response to this question, this chapter employs a critical literature methodological stance by using some of the specifications enshrined in the "critical interpretive synthetical" typology of the studies of Dixon-Woods, Cavers, Agarwal, Annandale, Arthur, Harvey, Hsu, Katbamna, Olsen, Smith, Riley, and Sutton.[8] Past relevant studies on the role of social media in the economic development were deconstructed and critically synthesized to ascertain whether they interrogate the socioeconomic realities of Africa. The importance of qualitative methodology is that it contains elements from grounded theory that empowers data to be used at the end of a given study to generate theoretical postulations according to Dixon-Woods et al.[8]

4.2 THE OLD VERSUS THE NEW

It is often said that the new is the old in disguise because there are simply micro changes that do not really affect the macro concept of what has been changed. That may be true, to an extent, with human communication. However, what is happening is that almost 65 years after the debut of global television, traditional forms of electronic communication (newspapers, radio, and TV) are now outpaced by mobile telephony (cell phones) and internet-driven smartphones and tablets. The most fascinating and intriguing aspect is that not only are users enamored by the irresistible attractions of popular social media networking sites like Facebook, Twitter, Instagram, WhatsApp, Viber, Pinterest, and Snapchat but are also using African-created sites that function similarly like those in the West. In short, they simultaneously belong to both African created platforms and western platforms for information sharing.

Research has shown that social media (open source technology) platforms on the Internet are the new phenomenon, where information is transmitted between the citizens and their government.[3,2,9,17] What is even more promising is the fact that Africans created platforms, unique to each region and state such as Mxit, AfricaZone, BlueWorld, Kokoliko, NaijaPals etc., are emerging to cater the needs of students, women, prenatal health, medium-sized enterprises and network technology sharing needs for people on the African continent. With mobile banking burgeoning in East Africa through the use of mobile phones, there is hope for sustained socioeconomic development through mobile technology in the continent. Even in the political arena, publications by Langmia[10] on the use of YouTube as a medium for political electioneering in Kenya and Cameroon is testament to the fact that the future of political elections in Africa is going to witness a developmental phase as it will be carried out mostly on social media.

4.3 CONCEPTUAL DEFINITION OF SOCIAL MEDIA

Social-media communications powered by SNS are one of the various channels of communications that enable users from every nook and cranny to send and receive instant messages to loved ones and friends mostly using electronic devices like cell phones, smartphones, iPads, tablets, laptops, Macs, and personal computers. Scholars of social media have not yet agreed on a unified definition for social media,[4] because of the very nature of the changing phase of these media. Since 2005, with the creation of Myspace, as the rallying site for young people to meet and share information, today we have Facebook, Twitter, LinkedIn, WhatsApp, Viber, YouTube, Instagram, Snapchat, Pinterest, Tumblr, Flickr, Skype, and Google Hangout that have emerged on cyberspace. Some of these platforms have unique characteristics with respect to video sharing, audio, and texts. That is why a common definition that pertains to all of them can be tricky to conceptualize. Nevertheless, scholars do agree that they constitute a novel electronic form of communication between humans.[4,19]

These SNS have taken us by surprise. They have stormed the human-communicative sphere with new platforms, terminologies, character, and format of communications unbeknownst to mankind. As a result, we are pushed and pulled by multiple, sometimes interrelated, forces to either comply or be left out, and if one chooses the latter, one will certainly be counted as an outsider, who finds comfort in the old communicative media. Mostly, if not all countries, have joined the bandwagon and have restructured their human-communication structures to ally with the demands of social media. It is no longer a secret that President Obama has singled himself out as the social media 21st-century leader as he used to send weekly communicative messages to the American people by compressed audio and video feeds on YouTube. Many other press releases from his government were sent out using Twitter, Facebook, and YouTube. In fact, as the author was writing this chapter, he paused and spent about 3 min scrolling over YouTube page with videos of the President Obama's speeches, interviews, and basketball games. Comments from users are equally posted on the site. This is unprecedented, and has, if not already, changed governance and public interests globally and notably in the African continent where electronic communications have begun to make headway.

African governments have begun to recognize the importance of social media, especially as they could help fight many social, economic, and educational challenges facing the continent.[18] It is not uncommon to find new Ministries of Information Communication and Technology springing up in most African Countries. In fact, at the 2013 World Summit for Information Society, most of the African delegations constituted Ministers of new media communications. With new terrorist threats being posed by Boko Haram, and other border-related problems, including, but not limited to domestic problems, Governments, and NGOs; independent agencies doing business in whatever form in Africa in either Anglophone, Francophone or Lusophone Africa need social media platforms for transmitting and sharing quick requisite information that will help them navigate the terrain in the continent. Given that Africa enjoys 80% cell phone subscription rate and some of them powered by the Internet, there is hope for socioeconomic and cultural development in the continent just by carrying out data transmissions from one to many.

4.4 AFRICAN SOCIAL MEDIA ECOSYSTEMS

A plethora of social networks—modeled from the western created ones—have mushroomed in Africa just in the last 5 years. They include: "Camerborn" (Cameroonian social network), "NaijaPal" (Nigerian social network), "Mxit" (Mobile and Social network app based in South Africa), "Koko Liko" (Ghana's social network), Kenya Social Network (Kenya's LinkedIn version), etc., representing the various regions, states and ethnicities. These networks build social cohesion and connect people with similar interests. Also, professional networks like Kenya's version of LinkedIn are vital to human capital development within and beyond the continent.

4.4.1 MXIT

Mxit is primarily the South Africa-based social media platform that has transformed lives of people in the region and beyond. One of the most powerful roles of this platform is that it has partnered with other international online and offline organizations to bring counseling services, free daily pregnancy advice from reputable personalities to expectant mothers. It has also partnered with Facebook, Pinterest, Twitter, and has uploaded videos on YouTube about the benefits of belonging to Mxit. Many young people involved in drugs and other related negative activities have been able to subscribe and receive free counseling services from people able to get them out of the shadows. Educationally, Mxit has inspired users by letting them have access to educational applications (apps) that will make difficult subjects, like mathematics, seem easy for young students.

By partnering with UNICEF, they have actually struck a milestone with access to experts that will offer free services to educational and child care developmental challenges.

Mxit is also helpful in promoting company brands and promoting products. As stated on their site, they have had a reach of over 9 million users and these users are not only confined to South Africa. They have users in Nigeria, India, and Indonesia. So, in essence, people's lives have been improved by accessing Mxit according to the YouTube video uploaded on their site that has shown some statistics with respect to users and visitors. A study conducted by Bidwell and his colleagues on the cell phone in South Africa, they realized that among other uses that Mxit systems where users can use the "Call Back" free service offered by the carrier is important. A good example from their study is when someone uses the costless call back service to send an urgent message of sickness, and the caller gets quick feedback from the caller. In another study, Makoe[12] examined positive educational benefits that can accrue to distance learners who use Mxit. It goes a long way to facilitate collaboration between learners stationed in different cities of countries. Of course, there are other downsides to Mxit as portrayed in the study by Chigona et al.[5] In their study, they portray how Mxit could serve as a medium for encouraging delinquency among youths, as well as creating a breeding ground for pedophilia. The argument on this view is the same with any other social media platform that are populated by youths. Like regular offline media, there have to be measures taken by the providers and governments to ensure security and any breach could be liable to standard legal prosecution.

4.4.2 NAIJAPAL

NaijaPal is mostly popular among Nigerians citizens. According to Akpan et al.,[1] NaijaPal is one of Africa's largest online social networking and entertainment communities, which enables users to meet new people. It provides updates on Nigerian entertainment and metro news, GISTs, opinions, latest music, Nollywood movies, confessions, and interactive discussion forum. NaijaPal came into existence in 2007 (p. 359).[1]

This forum like other SNS helps in social development of youths in Uyo, Nigeria. Information acquisition and interdependent social development is key to resolving youths' unemployment and delinquencies, which is a result of idleness. The fact that they can log on to a site where mostly people of their language, religion, and cultural community are present provides a psychological relief and sense of belongingness. Socially, this is very important given that youths are the ones that populate the SNS more than any other demographics. At first, it was problematic to have everyone consume a single medium (radio) during the era of traditional media. Governments often controlled the traditional media in Africa, and propaganda programs that populated them from print to video. Nowadays, especially young people, have gravitated toward less censored and less controlled social media networking sites where they are free to set and control their own agenda from politics to culture.

4.4.3 BLUEWORLD

BlueWorld is a social network created by a media guru called Charl Norman mainly for the South African youths. Just like Facebook in the West, subscribers can create profiles of their own and upload pictures that they can share with friends and acquaintances. They are also allowed by the creator to create blogs and exchange text messages as well with their interlocutors. Increasingly, people are learning about BlueWorld. This year alone, the membership or subscriber rate has gone up to 137,000 members.

4.5 CHALLENGES FOR AFRICAN SOCIAL MEDIA

4.5.1 NEW MEDIA TECHNOLOGY CURRENCY AND SOCIOCULTURAL POVERTY

The spade pouring the sands of new media technology unto the band-wagon of Africa's postindependence train has been overwhelming. It has appeared from nowhere, like a thief in the night. Almost like the replica of imperialist conquests of the continent that came unannounced, it has taken some people by surprise. There are concerns about the future of forms of new media of communications, overburdening the traditional forms in a once traditionally oriented and cohesive society. True to an extent, the urban cities in Africa have been able to embrace this newcomer, as they have constantly done during and after the 1884 Berlin conference, that partitioned their continent. However, the rural, especially the remote rural parts of Africa with inhabitants like the elderly, the custodians of culture, Rural English/French/Portuguese illiterate women, the small people and Borroros of Cameroon, the Bushmen of South Africa, or the Maasai of Kenya, who have known relative quiet simple down-to-earth life patterns void of cacophonous Western "modern" lifestyle are seeing their cultures not only eroding but also threatening their own very livelihood.

The so-called new media currency brought to the continent without their consent, are buying their culture and traditions with or without their much-needed approval. Today in Africa, banking is carried out electronically, and schools have begun offering online classes using complex technology from abroad, and the fight for closing the digital divide gap, which is pushing IT organizations to venture and introduce these ICTs to this technologically illiterate group of people and their likes in other parts of Africa. Social-media communication platforms abound in cities and the government and the citizens are now somewhat involved in some kind of synchronous/asynchronous-communication transaction. Therefore, no one has to be left behind as the ITU has decreed (see World Summit for Information Society communique, WSIS).

The question that continues to haunt us as we indulge in this type of activity is whether these groups of people have been prepared long enough for such a drastic change in their lives. Rather, whether the non-Western illiterate Africans were sociopsychologically prepared to welcome this visitor. The same can be said about other disadvantaged people in our communities like the blind, the deaf, and the dumb. But the latter group at least had years of Western educational facilities aiding their communication and educational skills since colonization. It is not uncommon, at least in some parts of African cities to find schools for the blind and the handicap.

Therefore, it is quite feasible that integrating them into the scientific rigors of new media technology cannot be a daunting task. But that of the other unrepresented groups aforementioned, those that have chosen to remain with near rudimentary ways of life and living in close proximity to animals and plants, did not appear in the radar of merchants in western capitals that designed new technological tools of communications. They have become the by-product of Western technology that seems to engulf their entire ways of life. Foreign norms and culture in Africa sounds like a welcome relief even if it means excavating old backward cultures and traditions like those of the small people, Bushmen, Borroros, and

Maasai. Be that as it may, it is better to aim at hybridization at the early stage. It is often said it is not easy to be left handed at old age; meaning that physical and cognitive abilities of the mentally challenged are to be taken into consideration when introducing new human communication transactions.

4.6 TECHNOLOGICAL DEPENDENCY

Technological determinism, some have argued should propel humans to crave for new creative ways of doing things. But that has not always worked for Africa. Any new requirements, education, training, testing, adaptation, and implementation were neither possible with colonization nor with industrial nor Internet revolution in Africa. Western consumerism has gripped the continent. It is as if we shot ourselves on the foot and turned a blind and angry eye at what our ancestors left for us. Their technology was backward, so says the silent voice in our heads. To make our lives more prosperous, we have to completely embrace what comes from the West.

Ali A Mazrui, the Africanist matador of political thought, confesses that Africa has acquired the means of destruction instead of means of production in their contact with the West. However, according to him, it does not mean that they did not try to knock the door of the West for help. However, each time they wanted to learn technology by sending experts to industrialized nations in the early 60s, their requests were denied.[13] The West preferred to ship finished technological products to them, at cut-throat prices, even when they exported the raw materials to them in the first place. This became a burden since they were operating from the wrong side of the economic equation. Impoverishment continued unabated, and had been since independence. However, more and more African experts have privately and voluntarily studied advanced technological skills in the West but poverty, malnutrition, diseases, unemployment, youth delinquencies still continue to haunt the continent. Part of the reason is that mass exodus of Africans to fetch greener pastures in the West has accelerated and brain drain for Africa is fast becoming a brain gain phenomenon for the West. As a result, Africa has not ceased from being a highly consumer continent and the dumping ground for abandoned goods and services in Europe, America, and China.

It is generally true, even if some may argue otherwise, that when Europe and America sneezes, Africa catches a cold, to parody the French aphorism. Their new ways of communication using the Internet to reach people all over the world far supersede what Africans have designed and so we have to fall on the line behind them and devote our entire economic sustenance to their mercy. Since we failed to have them accept the offer to train our experts to return to the continent and make it a better place, technological quests for all its glories must be our daily morning prayer for socioeconomic development and educational empowerment for all and sundry in the continent. Migration cannot be our long-term goal. Those who have gone there on their own volition and have been attracted by their economic attractions and modern civilizations have not yet returned to the continent or at best have preferred to send remittances to their economic ailing relatives back home using social media platforms, like WhatsApp.

This has been the forlorn economic situation regarding the relation between Africa and the West after colonization. The new phenomenon commonly referred to as ICTs are shaping the interpersonal, intergroup, and mass communication in the continent. Schools, government services, NGOs, small and medium-sized enterprises, and cultural and traditional entities like Fondoms, Chiefdoms, and Kingdoms are integrating their services with ICTs[6] in World Wide Web or social media.

To achieve these objectives, we must collectively embrace social media communication because that is the future of global communications. Social media, like any new media platform, take time to create an impact. User-ship can sometimes be complicated. There is no formal schooling in the use of these burgeoning platforms and SNS. There is the growing, and vexing presence of the digital divide in Africa, where most people rely on cyber cafes to access the Internet and those who have cell phones may not have wireless capabilities to log on to the SNS like WhatsApp, Viber, and Facebook, which is increasingly growing in the continent. The situation is even direr with respect to the non-Western educated folks in the rural, as well as the urban centers in the continent. To adapt the Western-made communication, forms have always been a daunting task to this group of people. Those Africans trained in Western style education have relative ease with their use because they read and obeyed instructions on their smart phones and computers. Imagine people with limited Western education, trying to navigate those complex social media apps and networking sites on the Internet. Just remembering passwords is a huge challenge because you have to constantly change them to avoid hacking, spams, and worms.

There are other numerous challenges that accompany this new technology. Language barriers between nations are impeding the sharing of information. There are Anglophone, Francophone Arabic, Amharic, and Lusophone African countries. Additionally, some of these countries also have native languages that are not easy to read, write, and interpret from mobile devices using European keyboards and icons. Digital and mobile divides, from a language point of view whereby only a select few are educated in Western languages do utilize all the features on mobile devices.

Gender and age gap with respect to ICT is another challenge. Some universities, especially in Tropical Africa have begun inculcating cyber knowledge in their campuses. Recent initiatives to spread the use of ICTs in Africa spearheaded by Professor Victor Mbarika has seen ICT institutions established in Uganda, Nigeria, and Cameroon with the intent of training future Africans through virtual means of communication using open source systems like WebEx technology and other Internet-driven platforms like Google Hangout and Skype to teach students. We have students from all over Africa that are taking courses virtually from that university.

Another challenge is regulation. Regulatory concerns that address both the traditional and new media platforms are one of the problems confronting most countries in Africa today. At first, regulation was the preoccupation of traditional media like radio, newspapers, and television. With the advent of convergence; it is not an easy step to take when examining what to regulate, how to regulate, and what not to regulate as new forms of digital media communications through apps and software are developed at a much faster pace. New media communication has now become like the Hydra, whose head appear and reappear in the mobile devices we buy every day.

4.7 SYNTHESIS AND SUMMARY

The various scholars of social media have argued in support and against the role of new media as an agent for promoting socioeconomic development, especially in Africa. They have conducted researches that support their various cases. When Yost and Fan[19] and Carr and Hayes[4] brilliantly articulated the role of new media platforms as the novel form of communication for humans, they are in fact arguing for the inevitable omnipresence of social media that have introduced a new form of message transaction for development. Their thesis has been the driving force behind the research question for this study. This study set out to find out how related literature on social media address socioeconomic imperatives of Africa. Based on the works of Makoe[12] online distance learning spearheaded by the force of ICTs has been a huge success for Africa because, for once, the continent is able to have citizens from all the regions gain and transmit knowledge across the world by using the Internet technology. This could be the way to resolve the brain-drain saga that has rocked the intelligentsia on the continent at the same time stemming the tide of mass immigration to the Western world. However, at the same time, we are increasingly facing the digital-divide woes, haunting users on the continent whereby not everyone has reliable access to the Internet or educated on how they function. Still, with regards to the research questions, there are other scholars like Chigona et al.[5] who have presented the negative side of social media especially as it relates to the African-driven platforms like Mxit originating from South Africa. This South African-generated user platform is mostly used by youths and as such they cater to the needs of the younger population. But as Chigona et al. [5] have pointed out this medium tends to encourage youth delinquencies and pedophilia among the youths. Be that as it may, there is no social-media platform used for socioeconomic development of a given community that does not have its downside.

The fact that Mxit has become so popular and attracting, users everyday goes to buttress the fact that youths are finding some information gleaned from this site useful to make their lives socially and economically better. This is definitely the case with the study carried out by Akpan et al.[1] This study showed the benefits users get when they log on to NaijaPal, a social media platform for Nigerians at home and abroad. One of the benefits they rip from the site is knowledge and information about entertainment updates on music, film, TV, and people confessing on things that matter in their lives. In conclusion, social media platform in this case has been adopted to fit the life style of those in given communities. True, Facebook, WhatsApp, Twitter, LinkedIn, and Instagram are products that are created and managed by the West but at the same time if non-Western users, in this case Africans can Africanize them through subject-matter content and language, as the case might be; then, it is possible to conclude that social media are a hybridized form of communication that can be used to uplift the living standards of its users.

KEYWORDS

- **Africa**
- **communicative arena**
- **G-20s**
- **International Telecommunication Union**
- **socioeconomic**
- **social media platforms**
- **social networking**

REFERENCES

1. Akpan, U.; Akwaowo, E.; Senam, N. Uses and Gratification of Social Networking Websites Among Youths in Uyo, Nigeria. *Int. J. Asian Soc. Sci.* **2013,** *3* (2), 353–369.
2. Axford, B. Talk About a Revolution: Social Media and the MENA Uprisings. *Globalizations* **2011,** *8 (5),* 681–686.
3. Adeiza, M. The Trouble with Social Media in Africa. *New Afr.* 2013, *526,* 78–79.
4. Carr, C.; Hayes, R. Social Media: Defining, Developing, and Divining. *Atlantic J. Commun.* **2015,** *23* (1), 46–65.
5. Chigona, W.; Chigona, A.; Ngqokelela, B.; Mpofu, S. Mxit: Uses, Perceptions and Self-Justifications. *J. Inform. Technol. Org.* **2009,** *4* (1),1–16.
6. De Bruijn, M.; Brinkman, I. Communicating Africa: Researching Mobilekin Communities Communication Technologies, and Social Transformation in Angola and Cameroon. *Autrepart* 2011, *1* (57–58), 41–57.
7. Dijck, J. V. *The Culture of Connectivity: A Critical History of Social Media;* Oxford University Press: NewYork, 2013; p 240.
8. Dixon-Woods, M.; Cavers, D.; Agarwal, S.; Annandale. E.; Arthur, A.; Harvey, J.; Hsu, R.; Katbamna, S.; Olsen, R.; Smith, L.; Riley, R.; Sutton, A. J. Conducting a Critical Interpretive Synthesis of the Literature on Access to Healthcare by Vulnerable Groups. *BMC Med. Res. Methodol.* **2006,** *6* (35), 1–13. https://lra.le.ac.uk/bitstream/2381/61/3/critical%20interpretive%20BMC%20Med%20Res%20Meth.pdf (accessed Jun 22, 2018).

9. Klein, A. Participatory Media Impact on the Arab Spring. In *Social Media: Pedagogy and Practice*; Langmia, K.; Tyree, T.;O'Brien, P.;Sturgis, I. Eds.; University Press of America: Lanham, MD, 2014; pp 182–203.
10. Langmia, K.; Tyree, T.; O'brien, P.; Sturgis, I. Eds. *Social Media: Pedagogy and Practice*; University Press of America: Lanham, MD; 2014; p. 275.
11. Lutterell, R. *Social Media: How to Engage, Share, and Connect;* Rowman and Littlefield: Lanham, MD, 2015; p 260.
12. Makoe, M. Exploring the Use of Mxit: A Cell Phone Social Network to Facilitate Learning in Distance Education. *Open Learn.* **2010,** *25*(3), 251–257.
13. Mazrui, A. *The Africans: The Triple Heritage;* Little Brown and Co.: Boston, 1986; p 328.
14. McLuhan, M. *Understanding the Media: The Extension of Man;* MIT Press: Massachusetts, 1994; p. 392.
15. Newman, N.; Dutton, W. H.; Blank, G. Social Media and the News: Implications for Press and Society. In *Society and the Internet: How Networks of Information and Communication Are Changing Our Lives;* Graham, M.; Dutton, W. Eds.; Oxford University Press: New York, 2014; pp 135–148.
16. Bidwell, N. J.; Lamas, M.; Marsden, G.; Dlutu, B.; Jones, M.; Tucker, B.; Robinson, S. Please Call ME.NU 4EVER: Callback and Social Media Sharing in Rural Africa. *Proc. IWIPS* **2011,** *11,* 117–138.
17. Terblanche, N. You Cannot Run or Hide From Social Media-Ask a Politician. *J. Public Aff.* **2011,** *11* (3), 156–167.
18. Thigo, P. People, Technology and Spaces: Towards a New Generation of Social Movements. *J. Contemp. Afr. Stud.* **2013,** *31* (2), 255–264
19. Yost, H.; Fan, S. Social Media Technologies for Collaboration and Communication: Perceptions of Childcare Professionals and Families. *J. Early Child.* **2014,** *39* (2), 36–41.

CHAPTER 5

Mobile App-Based Expert System in Uganda: Combating Pests and Diseases in Tomatoes

EMMANUEL EILU

ABSTRACT

Kayunga district is among the leading producers of tomatoes in Uganda. One of the reasons of low tomato production is rampant spread of pests and diseases, such as the Early Blight (*Alternaria solani*), Late Blights (*Phytophthora infestans*), and Bacterial wilt (*Ralstonia solanacearum*), and pests such as the broad mite (*Polyphagotarsonemus latus*), thrips (*Frankliniella occidentalis*), aphids (*Acyrthosiphon pisum*), and American bollworms (*Helicoverpa armigera*). Also, the farmers are not in a position to accurately diagnose and prescribe the right control measures. In this chapter, author has developed a mobile application that runs on any smartphone with android platform, and it allows a farmer to take sample pictures of parts of the plant that have shown signs of infection, namely, tomato fruit, leaf, or stem. The farmer then uploads the picture on the application for diagnosis and prescription.

5.1 INTRODUCTION

The gross domestic product (GDP) in Sub-Saharan African (SSA) countries in recent years has grown from 5% to 6% per year.[18] This GDP growth rate is among the highest in the world. Food and agricultural production in some SSA countries have also grown on a per capita basis.[18] Despite this remarkable economic growth and development, the agricultural sector has not witnessed a significant transformation in majority of SSA countries, it has been slow and the growth has been rather sluggish.[5] It is important to note that African farming industry has a lot of potential to grow, given the abundant availability of rainfall and land.[8] However, agricultural productivity is still extremely low, agricultural mechanization is feeble and declining, and the state of the agribusiness industry is still nascent.[5] As a result of low agricultural productivity, there has been a growing concern over poverty and food insecurity in the SSA region.[6,8] As already witnessed in the last decade, the agriculture sector in many SSA countries has been experiencing prolonged droughts and/or floods during El Nino events.[8]

Reports released on the state of food insecurity and malnutrition in SSA observed that more than a quarter of the region's population is suffering from hunger.[11] The report further revealed that 26% of the population in SSA has suffered from severe food insecurity in 2014–2015.[11] As a result, most of Africa's population has been impoverished because of the inability to harness the readily available agricultural resources, such as land and water for economic development and better livelihoods.[5] Besides drought and flooding, crop pests and diseases are responsible for the considerably low level of agricultural production. Research has shown a significant relationship between pests and diseases and crop sustainability in several areas of Kenya, Nigeria, and other regions of Africa.[9]

The research shows that the outbreak of various pest and diseases has led to a sharp decline in crop production and therefore putting the entire region on a food scarcity alert.[9] According to 2007 FAO report,[7] transboundary plant pests and diseases are main causes of pest and disease outbreak in the SSA region. These outbreaks not only cause food scarcity and famine but they also negatively affect pastures for the animal and the livelihoods of millions of people. Pests such as earthworms, fruit flies, and locusts are most common, while cassava diseases and wheat rusts disease are among the most destructive transboundary diseases.[7]

Due to the significance of the agricultural sector, which is the backbone of many economies in SSA and a major sector that contributes to the livelihoods of the poor, the World Bank advocates to place the agricultural sector at the center of the development if extreme poverty and hunger is to be reduced by half.[5] There is a need to boost agricultural production in SSA by using effective pest and diseases control measures, and proving timely agricultural information to farmers. One of the fastest and most effective ways of doing this is through the strategic use of Information and Communication Technologies (ICTs).

There is need to strategically deploy ICT in the agricultural sector, which by far and large is the biggest sector in most African countries in the Sub-Saharan region. This would offer a good opportunity for growth in the agricultural sector and a stop to the rampant food shortages in this part of the world.[27] The use of ICTs offers a significant impact on agriculture and has the potential of realizing the vision of Africa as a transformed continent by 2023.[14] The world is transforming fast, and ICT is at the forefront of this transformation process. It is therefore important that SSA, in general, quickly embraces the use of ICT in all the sectors of the economy including agriculture.[10]

There is a need to deploy ICTs and more particularly mobile phones in the agricultural sectors. ICTs, particularly, mobile technology is increasingly being adopted as the technology to facilitate the delivery of extension services to rural farmers.[27] Mobile phone coverage has been growing steadily from 10% penetration in 1999 to over 41% in 2008. It is expected to reach 50% penetration in 2020 over 500 million subscribers, from which 400 million are smartphone users.[3,21] A country like Uganda, for example, has a mobile phone penetration level of 100%. The country has over 2300 base stations and 19 million mobile phone subscribers.[21]

Therefore, because of the widespread mobile technologies, success has been registered in some SSA countries where the use of mobile devices has been deployed to transform the agricultural sector. In Kenya, Senegal, Niger, Ghana, and Uganda, mobile phones messages have widely been used by government and other organizations to educate farmer on how to improve crop yield, monitor fertilizer distribution, market farmer's produce, and access technical guidance.[3,15] Jere[10] observes that other applications of mobile technologies in agriculture include data collection, data analysis, geospatial applications, and decision-support and knowledge-based systems. The use of mobile phones in the agricultural sector is a panacea to improved yield.

5.2 OBJECTIVES OF THIS STUDY

As discussed earlier, one of the major problems affecting agricultural production in many SSA countries are pests and diseases, and also one of the solutions for addressing the pest and disease epidemic in the agricultural sector is the strategic use of ICT. Therefore, the main discussion for this chapter is on the use of mobile technologies to address the disease and pest problem in the agricultural sector in SSA countries. This chapter focuses on tomatoes as one of the most commonly grown vegetables in many SSA countries as a case study. This chapter also makes use of artificial intelligence (AI) as a technology for diagnosing the different pests and diseases that affect tomatoes in a typical SSA country like Uganda. The entire work proposes a mobile phone-based application that enables a farmer to accurately diagnose and prescribe correct pest and disease control measures on the common tomato pests and diseases. Pests include the broad mite, thrips, aphids, and American bollworms. Diseases include early and late blights and bacterial wilt.[2] However, in this chapter, author has discussed the success of the first phase of an ongoing work: diagnosing and prescribing correct control measures for the early blight.

5.3 CURRENT STATUS OF TOMATO CULTIVATION IN UGANDA: A CASE STUDY

As far back as 2003, FAO had identified tomato as the most important vegetable in the tropics with 4 million ha planted yearly worldwide.[8] Particularly in East Africa, tomato farming is practiced in Uganda, Kenya, and Tanzania.[26] Farmers in the East African region grow a variety of tomato species with most common being Italian processing varieties like: the Rio Grande, Roma, and Money Maker. They are preferred because they are high yielding, disease resistant, and have a longer shelf life than other varieties.[26] Tomato is among the most important high-value horticultural crops in Uganda offering opportunities for employment and access to education, food security, and health care to the growers.[12] The leading tomato production areas in Uganda are Mukono, Kayunga, and Mubende districts.[26] The commonly grown tomato commercial varieties include: the Moneymaker, Heinz, Roma, Marglobe, and local (Nganda). However, demand for tomato produce in Uganda often outstrips supply due to production challenges. Karungi et al.[12] and Akemn et al.[2] observed that there are two major challenges to tomato farming in these three districts of Mukono, Kayunga, and Mubende, such as:

- *Pests and diseases:* The most common tomato diseases in Uganda are: Bacterial wilt (*Ralstonia solanacearum*) and the early and late blights (*Alternaria solani* and *Phytophthora infestans*). While the major pests include: The broad mite (*Polyphagotaronemus latus*), thrips (*Frankinnela occidentalis*) aphids (*Acyrthosiphon pisum*), and American bollworms (*Helicoverpa armigera*)[2]. The blights and bacterial wilts[2] are the most destructive and causes almost 75% and 100% tomato yield loss, respectively if they are not well managed. It mainly affects the fruits, leaves, and stems.

- *Incorrect disease and pest diagnosis*: This leads to wrong pest and disease control measures. Maerere et al.[13] observes that the usage of wrong pesticides to control tomato insects and diseases in many developing countries is closely linked to the wrong diagnosis. This has led to high cost of production, low yields, and the potential for health and environmental risks associated with pesticides. Akemn et al.[2] observed that mixing and safe use of pesticides is ranked as the third most important constraint to tomato growers in Uganda.

In order to maintain the high yields, the modern farmer consults with agricultural experts in case there is a need for expert assistance. Unfortunately, Prasa and Babu[17] observed that this kind of assistance may not be readily available in many cases. Therefore, there is need to use technology to assist farmers who need expert advice. The use of expert systems (ESs) as a potential solution is recommended[17]. AI is a best solution to these agricultural challenges.[16]

5.4 AI/EXPERT SYSTEM

Introduced in the 1950s, the term AI or ESs can be defined as "*the science and engineering of making intelligent machines.*"[4]. These software use programmed expert knowledge. They give feedback to the user after a query session with the user.[1] In agriculture, expert knowledge in individual disciplines, such as plant entomology, pathology, agricultural meteorology, and horticulture are programmed into a framework that provides best solutions to farmer's on-site needs.[17]

AI is a relatively new field. Functioning like a human brain, this intelligent machine can identify problems and provide solutions effectively, and are widely been used in many areas of human life.[16] This technology is currently being applied in mobile phone technology. Intelligent mobile phone applications have been designed and developed to offer the much-needed expert knowledge in various fields including agriculture. Building an expert system on a widely used platform like mobile phones is cheaper and has better potential to provide agricultural solutions.

5.5 METHODOLOGY

5.5.1 *CASE STUDY: KAYUNGA DISTRICT, KAYONZA SUBCOUNTY, UGANDA*

This study was conducted in Kayonza Subcounty in Kayunga district of Uganda. According to Karungi et al.,[12] Kayunga District in the eastern part of Uganda is the leading tomato grower region in the country. Stakeholders are: Agricultural officer, tomato farmers, veterinary officer, and programmers/researchers. Interviews, questionnaires, and observation tools were used to collect data. This provided functional requirements of the mobile application. The agriculture officer and the veterinary officer were able to provide the relevant information on the early blight, particularly on the signs and treatment of the early blight. This useful information was then used to develop the diagnosis and prescription system.

5.6 DEVELOPMENT OF MOBILE APPLICATION DIAGNOSIS AND PRESCRIPTION TOOL

To develop the mobile application diagnosis and prescription the tool, author used: NumPy, Matlab, Flask Restful, Sklearn, Opencv, KNN classification, and Python as the main programming language.

5.7 PROCESS MODEL

The tomato disease diagnosis system uses the machine learning. In this case, the system is able to distinguish between a healthy and unhealthy tomato plant suffering from early blight disease. The model involves four distinct steps to accomplish its goal as discussed in the following sections.

5.7.1 *IMAGE CAPTURE*

Several images were taken from tomato gardens using a camera. The parts of the tomato images included tomato leaves, fruits, and stem for both plants that were infected with early bright and those were not. Categorization of captured images is as follows:

Group 1: Healthy plants
Group 2: Nonhealthy (infected with early blight)

5.7.2 IMAGE PREPROCESSING

The purpose of image preprocessing was to eliminate noise in tomato images, such as background clutters and nonuniform surface illumination there by giving a clear tomato image, which is later used in segmentation. A set of image before and after preprocessing are shown in Figures 5.1–5.6. This process retains only relevant parts of the leaf to be used for diagnosis by the system. The illustrations in these figures are for the tomato samples infected with early blight.

FIGURE 5.1 (See color insert.) Before preprocessing showing infected leaf.

FIGURE 5.2 (See color insert.) After preprocessing.

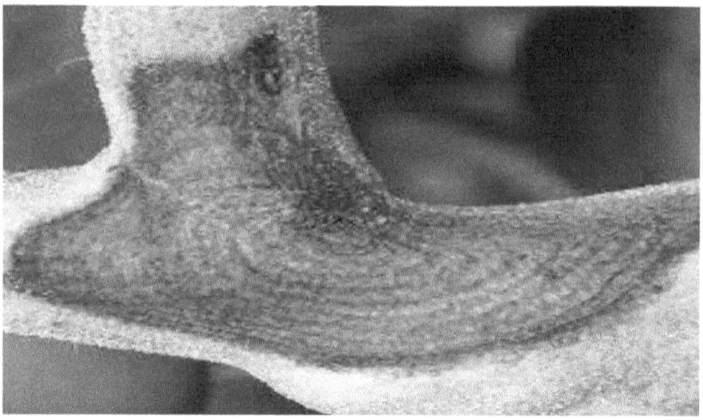

FIGURE 5.3 (See color insert.) Before preprocessing showing infected stem.

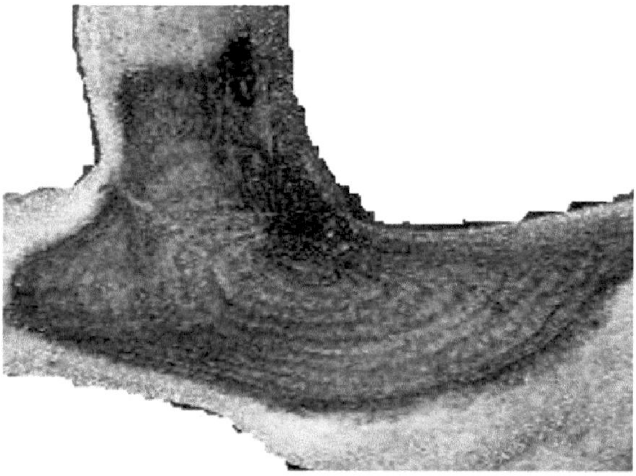

FIGURE 5.4 (See color insert.) After preprocessing.

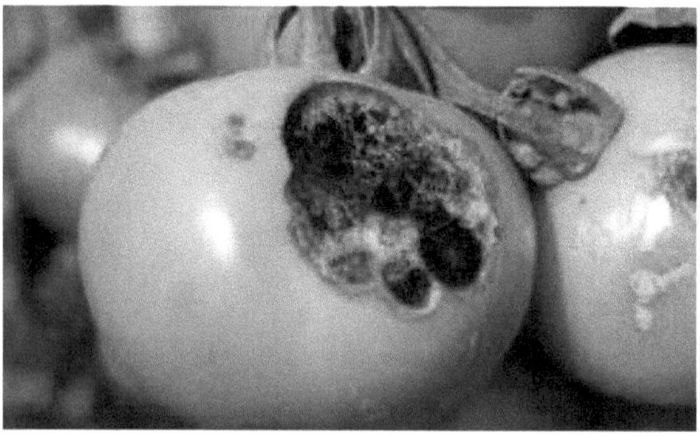

FIGURE 5.5 (See color insert.) Before preprocessing showing infected fruit.

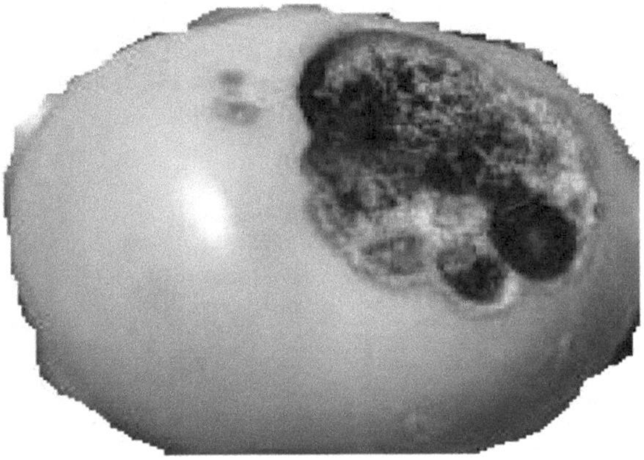

FIGURE 5.6 **(See color insert.)** After preprocessing.

5.7.3 SEGMENTATION

Image segmentation involves assigning a label to every pixel in an image whereby the pixels with the same label share certain characteristics. The images were divided into several segments making them easier and meaningful to analyze. This helped in locating the boundaries and objects (line and curves) in the image that were captured. The resultant outcome of image division was a set of segments or image contours extracted from the image of a relevant part of the leaf, fruit, and stem. Color, intensity, and texture were the characteristics represented in each pixel in a particular region (Fig. 5.7). The segmentation was done using Matlab.

FIGURE 5.7 Leaf, stem, and fruit after segmentation.

5.7.4 FEATURE EXTRACTION

This involves converting data gathered into the set of different features. The target features are identified and extracted. The extracted features will act as a representative of information from the input data in order to perform the desired study. The gray scale was used to extract special features in each of the stored image. A gray scale histogram below represents the number of pixels that have a gray scale in certain range (Fig. 5.8A–C).

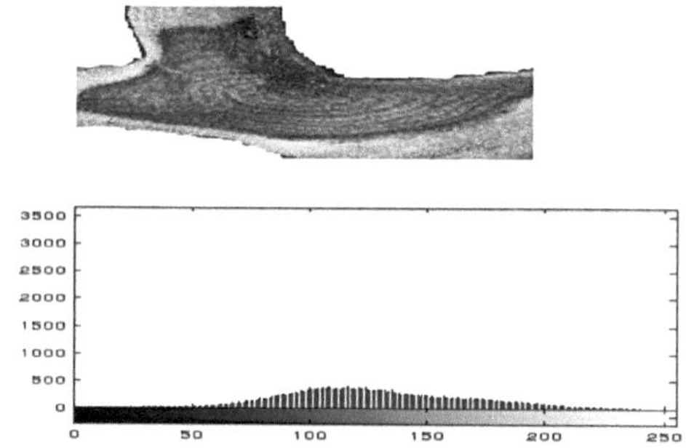

A

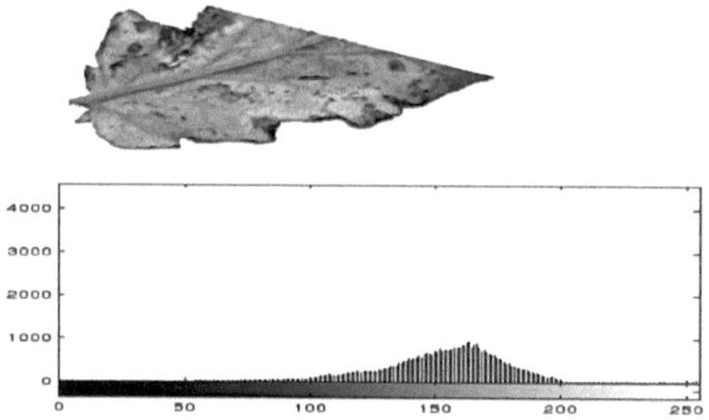

B

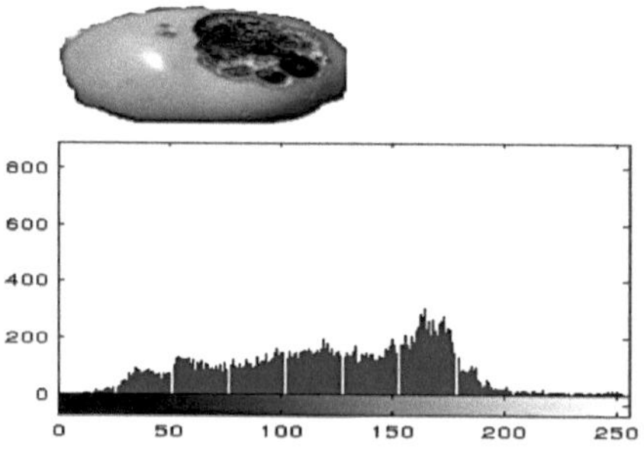

C

FIGURE 5.8 (See color insert.) Feature extraction.

5.7.5 CLASSIFICATION

This involves analyzing several properties of different images and categorizing them based on Naïve Bayes and the KNN. Each time a loop goes through the images, it appends a label to each image according the categorization. The entire process model is illustrated in Figure 5.9.

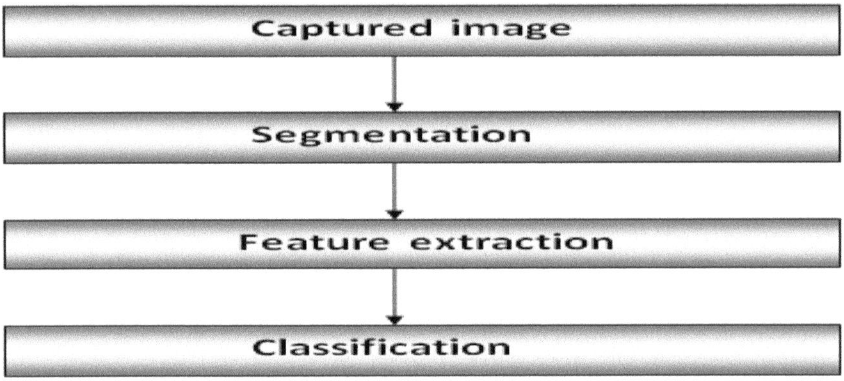

FIGURE 5.9 Process model.

5.7.6 TESTING THE CLASSIFIERS

Author compared the performance of the two classifiers (KNN and Naïve Bayes) used in terms of accuracy, sensitivity, and specificity, where accuracy depended on the images making up the training data set, that is, the less the number of images, the less accurate the system; the more the images making up the training data set, the more is the accuracy of the system.

True positive (TP): Images classified as infected with early blight when they are infected.

True negative (TN): Images classified as not infected with early blight when they are not infected.

False positives (FP): Images classified as infected when they are not infected by early blight.

False negative (FN): Images classified as not infected when they are actually infected with early blight.

$$\text{Accuracy} = [(TP + TN)]/[(TP + FP + TN + FN)] \quad (5.1)$$

From Table 5.1, Naïve is more accurate for small data sets while KNN performs very poorly due to very few neighbors. As the number of images increases, the two classifiers tend to perform in the same way.

TABLE 5.1 Testing the Classifiers.

No. of images	KNN	Naïve Bayes
1	42%	100
5	52%	100
10	65%	100
15	75%	75
20	85%	85

5.7.7 IMPLEMENTATION

The presentation of the results of the Tomato Disease Diagnosis System was analyzed in terms of the interfaces of the system and this includes activities of the farmer. Some of the activities include taking a sample picture of the infected tomato leaf, stem, or fruit and uploading the picture into the system. The implementation of the Tomato Disease Diagnosis System yielded several results shown in the screen shots of the interfaces in Figure 5.10.

The farmer activates the mobile application by touching and holding the launcher until it vibrates. The farmer clicks the select picture button to choose a picture of the tomato leaf, fruit, or branch from the gallery which was previously taken and stored in the gallery. Once the picture is located, the farmer selects the image to be tested and clicks on the "click to upload file" button.

5.8 RESULTS AND DISCUSSIONS

The mobile phone tomato disease and pest diagnosis application developed from the process model (Fig. 5.9) runs on any smartphone with Android platform. It allows a farmer who has a smartphone to take sample pictures of parts of the plant such as tomato fruit, leaf, and stem that have shown signs of infection (early blight). The farmer then uploads the picture on the application for diagnosis and prescription. Preliminary tests on the early blight indicate that the system is highly successful especially on fully developed signs of the disease on leaf, fruit, and stem. For best results, the user must carefully take the picture of the infected leaf, fruit, and stem. Out of the 30 samples that were tested for early blight, 26 samples were successful.

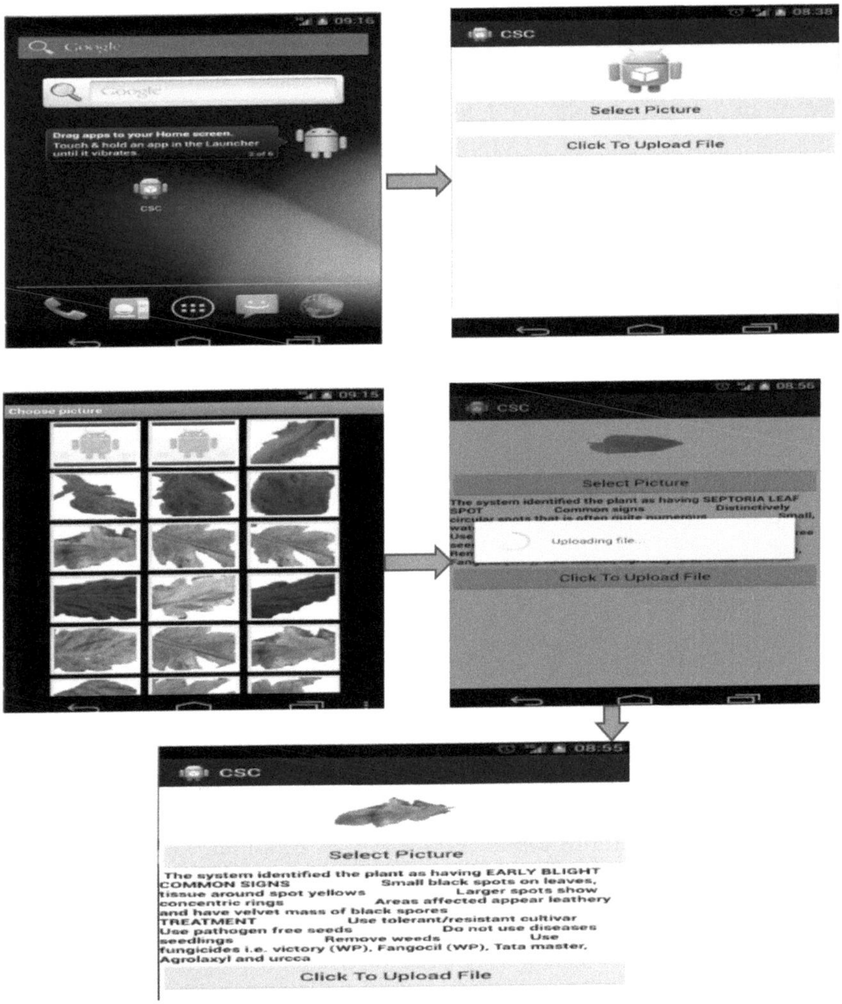

FIGURE 5.10 **(See color insert.)** The diagnosis process.

The research is currently in its first phase of diagnosing and prescribing control measures for early blight though it will roll out to late blight, bacterial wilt, and the other stated pest. As discussed earlier, incorrect disease and pest diagnosis in many cases leads to wrong pest and disease control measures. This is a very common phenomenon in Kayunga district in Eastern Uganda, which is one of the leading tomato producers in Uganda. Many farmers are illiterate and do not have knowledge of the different pests and diseases of tomato.

Shahzadi et al.[20] observed that many farmers in the remote areas are illiterate and do not have proper knowledge of dealing with pests and diseases. Farmers in Kayunga find it extremely hard to differentiate between the early blight and late blight in tomatoes. A number of tomato diseases display the same signs on the tomato leaf, fruit, or stem: it therefore is very difficult for rural farmers to differentiate between the diseases, hence ending up using the wrong pesticide.[20]

The application of incorrect pesticides to control tomato insects and diseases in many SSA countries is majorly linked to wrong diagnosis.[13] One of the solutions that can aid even illiterate farmers to accurately diagnose the different tomato pests and diseases is the use of ESs. Experts systems combine the knowledge of different experts in one application.[20] In this study, the introduction of an expert mobile application was relatively successful in diagnosing the early blight in tomatoes. A number of ESs have shown tremendous success.

5.9 SUMMARY

Kayunga district is among the leading producers of tomatoes in Uganda; the tomato production levels in this district have constantly been dropping due to two main reasons: (1) Rampant spread of pests and diseases and (2) wrong disease and pest diagnosis leading to inadequate control measures. In this chapter, author developed a mobile application that runs on any smartphone with Android platform. It allows a farmer with a smartphone to take sample pictures of parts of the plant that have shown signs of infection, namely, tomato fruit, leaf, or stem. The farmer then uploads the picture on the application for diagnosis and prescription. It should also be noted that this mobile application currently diagnoses and prescribes only for the early blight; however, the application has further been developed to improve its effectiveness and also to be able to diagnose other stated diseases and pests.

KEYWORDS

- **expert mobile app**
- **information and communication technology**
- **Sub-Saharan African**
- **tomato**
- **Uganda**

REFERENCES

1. Abdelhamid Y.; El-Helly, M. New Approach for Developing Diagnostic Expert Systems on Mobile Phones. *Commun. Information Sci. Manage. Eng.* **2013,** *3* (8), 374–387.
2. Akemn, M. C.; Kyamanywa, S.; Luther, G.; Ssekyewa, C.; Erbaugh, J. M.; Warren, H. Developing IPM Systems for Tomato in Central and Eastern Uganda. *IPM CRSP Sixth Annual Report*, The Ohio State University: Columbus, Ohio, 2000; pp. 117–121. 203.64.245.61/fulltext_pdf/EAM/1991-2000/eam0106.pdf (accessed Apr 26, 2017).
3. Aker, C. J.; Mbiti, I. M. Mobile Phones and Economic Development in Africa. *J. Econ. Perspect.* **2010,** *24* (3), 207–232 http://sites.tufts.edu/jennyaker/files/2010/09/aker_mobileafrica.pdf (accessed May 15, 2017).
4. Andresen, S. L.; McCarthy, J. Father of AI. *IEEE Intell. Syst. Mag.* **2002,** *17*, 84–85.
5. Chauvin, N. D.; Mulangu, F.; Guido Port, P. *Food Production and Consumption Trends in Sub-Saharan Africa: Prospects for the Transformation of the Agricultural Sector*; Working Chapter, February 2012, WP 2012-011; pp. 12–33. www.undp.org/content/dam/rba/.../Food%20Production%20and%20Consumption.pdf (accessed Jul 3, 2016).
6. Demeke, M.; Di Marcantonio, F.; Morales-Opazo, C. Understanding the Performance of Food Production in Sub-Saharan Africa and Its Implications for Food Security. *J. Dev. Agri. Econ.* **2013,** *5* (11), 425–443.
7. FAO (Food and Agricultural Organization). *Plant Pests and Diseases*, 2017; pp 1–7. http://www.fao.org/emergencies/emergency-types/plant-pests-and-diseases/en/ (accessed May 2, 2017)
8. FAO (Food and Agriculture Organization). *The Special Challenge for Sub-Saharan Africa*. 2009. www.fao.org/fileadmin/templates/wsfs/docs/Issues_chapters/HLEF2050_Africa.pdf (accessed Apr 26, 2017).

9. Goldman, A. Pest and Disease Hazards and Sustainability in African Agriculture. *J. Exp. Agri.* **2008,** *32* (2), 199–211.
10. Jere, T. N. *Unlocking the Promise of ICTs for Transforming Agriculture in Africa. NEPAD e-Africa Commission*; CSIR, Pretoria, South Africa, 2010; pp 16–23. http://knowledge.cta.int/Dossiers/S-T-Policy/ICKM-MIS-ICT/ICT-for-transforming-research-for-agricultural-and-rural-development/Feature-articles/Unlocking-the-Promise-of-ICTs-for-Transforming-Agriculture-in-Africa (accessed Apr 26, 2017).
11. Jungcurt, S. *Assessing Food Security in Africa.* 2017. http://sdg.iisd.org/news/fao-assesses-food-security-in-africa/(accessed Apr 15, 2017).
12. Karungi, J.; Ochwo-Ssemakula, M. K. N.; Sseruwagi, P.; Jurua, H.; Ahabwe, S. In *Developing Ecological Management Strategies for the Aphid Vectored Cucumber Mosaic Virus on Tomato in U*ganda. Third RUFORUM Biennial Meeting; Entebbe: Uganda; Sept 24–28, 2012, pp 18–21.
13. Maerere, A. P.; Sibuga, K. P.; Bulali, J. E. M.; Mwatawala, M. W.; Kovach, J.; Kyamanywa, S.; Mtui, H. D.; Erbaugh, M. Deriving Appropriate Pest Management Technologies for Smallholder Tomato (*Solanum lycopersicum* Mill.) Growers: A Case Study of Morogoro, Tanzania. *J. Anim. Plant Sci.* **2010,** *6* (3), 663–676.
14. Nosarka, C. Growing Use of ICT Is Helping Transform Agriculture in Africa; 2014. http://www.foodprocessingafrica.com/growing-use-of-ict-is-helping-transform-agriculture-in-africa/(accessed Apr 26, 2017).
15. Ogutu, J. How ICT Drives Kenya's Economic Growth. Standard Digital, 2015. https://www.standardmedia.co.ke/article/2000162611/how-ict-drives-kenya-s-economic-growth (accessed Apr 26, 2017).
16. Popa, C. Adoption of Artificial Intelligence in Agriculture. Bulletin UASVM Agriculture, Iaşi: România; **2011,** *68* (1), E-report; Electronic ISSN 1843-5386. http://journals.usamvcluj.ro/index.php/agriculture/article/viewFile/6454/5747 (accessed Apr 26, 2017).
17. Prasad, G. N. R.; Babu, V. A. A Study on Various Expert Systems in Agriculture. *Georgia. Elec. Sci. J.* **2006,** *4* (11), 22–32.
18. Rosen, S.; Shapouri, S. Factors Affecting Food Production Growth in Sub-Saharan Africa. 2012, *10* (3), E-article. https://www.ers.usda.gov/amber-waves/2012/september/factors-affecting-food-production/ (accessed Apr 26, 2017).
19. Ryan, P.; Tsunemine, K. Sub-Saharan Africa, Special Study Edition; Sept, 10, 2016; pp 21–25. https://www.marubeni.com/research/report/political_economy/global/data/24-Sub SaharanAfricaSpecialEdition.pdf (accessed May 2, 2017).
20. Shahzadi, M.; Tausif, R.; Ferzund, J.; Suryani, M. A. Internet of Things Based Expert System for Smart Agriculture. *(IJACSA) Int. J. Adv. Comp. Sci. Appl.* **2016,** 7 (9), 341–350.
21. UCC (Uganda Communication Commission). Post, Broadcasting and Telecommunications Market and Industry Report, 2015. http://www.ucc.co.ug/files/downloads/Q4-Market-Report-for-October-December-2015.pdf (accessed May 15, 2017).
22. USAID. The Fresh Fruit and Vegetable Markets of East Africa. *An Assessment of Regional Value Chain Actors, Activities and Constraints in Kenya, Tanzania and Uganda.* ICTs for Agriculture in Africa: Uganda; 2013; eTransform AF.

CHAPTER 6

Role of Mobile Phone Technology in Combating Illegal Timber Logging in Africa: A Review

EMMANUEL EILU

ABSTRACT

More than 15 million ha of tropical rainforests in Africa are being destroyed so as to pave way for small- and large-scale agriculture, cattle rearing, firewood, and illegal logging for timber. As a result of high rate of deforestation, about 13 million tons of soil are washed yearly into the water bodies in Africa. Illegal logging contributes to about 70% of the forest destructive forces. If the current destruction pattern of rainforest is sustained, it is estimated that there will be no more forests left by 2050 in Africa, hence contributing to a park of devastating effects (such as climate change, loss of biodiversity, global warming, land degradation, desertification, drought, food security, and poverty). One major way to significantly reduce deforestation is to combat illegal logging. There are several ways of curbing illegal logging in Africa; however, many of these methods are quite expensive for the cash-strapped countries. Nevertheless, the use of technology, particularly mobile phones, can be an inexpensive way of curbing illegal logging, and currently, a handful of such projects are being piloted across Africa. However, very little has been published on how recent mobile phone innovations have been used to combat illegal logging in Africa. Using systematic review method, this chapter attempts to bridge this gap by exploring recent innovations in mobile phone technology used for curbing illegal logging in Africa.

6.1 INTRODUCTION

The continent of Africa is home to second largest blocks of tropical rainforest in the world after the rainforest of the Amazon Basin in South America. There are number of tropical rainforests in Africa and the largest is the Congo basin forest that is spread across 170 million ha of land.[10] The numerous rainforests stretch for about 4800 km from Liberia in West Africa through Ivory Coast, Ghana, Nigeria, the Cameroons, French Equatorial Africa, and the Belgian Congo, to the great lakes region in East Africa.[2] These different rainforests scattered all over Africa are very vital to the survival of the entire continent, that is, they balance the ecosystem, regulate both local and global climate patterns that support biodiversity, carbon sequestration, and timber and wood fuel production, which boost economic growth and development in Africa.[11]

However, Africa is currently witnessing a new wave of deforestation, which is sweeping across the entire continent destroying wildlife and incapacitating the resilience of its ecosystems to withstand the effects of climate change, especially in the area of food security, desertification, and global warming.[7] There are local variations in which tropical rainforests are being destroyed. For example, in countries like Nigeria, the Ivory Coast, Ghana, and Madagascar, rainforests are normally destroyed to pave way for agriculture, whereas in some parts of DR Congo, Congo Brazzaville, and Gabon, the rainforests are being logged for timber.[5] Statistics show that more than 15 million ha of tropical rainforests are destroyed to pave way for small- and large-scale agriculture, cattle rearing, firewood, and timber.[18]

According to the Africa Regional Review Report on Energy for Sustainable Development,[1] in some sub-Saharan African countries like Burundi, Rwanda, Mozambique, and Burkina Faso, wood fuel constitute over 87% of the energy used for heating. In Ghana, 70% of the wood fuel used for domestic heating comes from the rainforests, resulting into the destruction of over 20,000 ha each year, while in Uganda, 90% of the population lives in rural areas and their major activity is agriculture, this activity has made the rainforest shrink from 45% in 1890 of the country's surface area to 21% in 2000.[1,18] As a result of high rate of deforestation, estimates show that about 13 million tons of soil is washed yearly into the water bodies in Africa.[19] If the current destruction pattern of rainforest is sustained, it is estimated that there will be no more forests left by 2050 in Africa, hence contributing to a park of devastating effects such as climate change, loss of biodiversity, global warming, land degradation, desertification, drought, food security, and poverty.[19]

The preservation of tropical rain forests in Africa is now becoming a global issue. One of the ways to preserve the tropical rainforests in Africa is to fight illegal logging. Illegal logging can be defined as logging practices and activities in violation of national law.[10] The World Bank[20] estimates that about US$ 10 billion is lost due to illegal logging, and this has made governments loose almost $5 billion in revenue.

According to Blaser,[4] millions of ha of forested land are made bare without prior authorization. This activity is estimated to generate between $30 billion and $100 billion yearly representing between 10% and 30% of the total global wood market.[16] The 2014 assessment conducted by INTERPOL found that illegal logging for charcoal trade is a vital source of revenue for criminal organizations and militias in African countries like the Democratic Republic of the Congo, Uganda, Rwanda, Somalia, and Tanzania.[16]

However, there are several ways of curbing illegal logging, however, many of these methods that would help curb illegal logging in Africa are quite expensive for the cash-strapped countries. Nevertheless, the use of technology, particularly mobile phones, can be an inexpensive way of curbing illegal logging. Statistics show that there were about 386 million mobile phone subscribers in sub-Saharan Africa in 2008. However, this figure had doubled in 2015, and on average there was 41% mobile phone coverage in number of countries in the sub-Saharan region. Forecasts on mobile phone ownership in sub-Saharan Africa put it at nearly 50% ownership by 2020, with over 500 million mobile phone owners and three quarters of the phones will be smartphones.[13] Uganda, for example, has over seven giant competing telecommunication firms, and the total population with mobile phones stand at about 19 million.[17]

With over 2300 base stations erected all over the country, the mobile phone coverage in Uganda had reached 100% in December 2009.[17] The availability of mobile phones in sub-Saharan Africa has provided technological means of overcoming illegal timber logging in Africa. A handful of such projects are being piloted across Africa. However, very little has been published on recent innovations in mobile phone technology used for curbing illegal logging in Africa. This chapter, therefore, attempts to bridge this gap by exploring recent innovative ways of using mobile phone technology to curb illegal logging Africa. Two research questions were developed to guide this exploration, namely:

- What are the recent innovative ways of using mobile phone technology to curb illegal logging Africa?
- What are the challenges of deploying these innovations for curbing illegal logging in Africa?

6.2 METHODOLOGY

6.2.1 SYSTEMATIC REVIEW

This chapter adopted systematic review[8] as the main methodology. Systematic review involves a step-by-step comprehensive and detailed plan, following a given literature search strategy based on given research objectives or questions. The goal of the systematic search of literature is to minimize bias by identifying, evaluating, and fusing all relevant studies on a particular topic based on the research questions. Its strength lies in the use of knowledge base to find out what works and how it worked.[21]

This method has widely been used in health-related studies and the natural sciences in an attempt to find out what works and how it worked.[21] This chapter comes up with clear research questions, as already stated above, these questions have guided the structure of the chapter. A wide and impartial literature search was conducted. The literature search was guided by key research questions. Keywords were identified from the research questions. The review covered: conference proceedings, journals, reports, websites, and books. The selection of chapters was based on how each chapter comprehensively addressed each of the research questions and particularly the keywords.

6.3 RESULTS: INNOVATIVE WAYS OF USING MOBILE PHONE TECHNOLOGY: CURRENT STATUS

This section presents results of recent innovative ways of using and deploying mobile phone technology to curb illegal logging in Africa.

6.3.1 FORESTLINK: CAMEROONIAN

Dunn[9] observes that, in response to illegal logging, the Cameroonian government partnered with number of development partners such as the Department of International Development (DFID), the Waterloo Foundation, and The Rainforest Fund FODER (Forest and Rural Development) to deploy a program called ForestLink. The system allows surrounding forest dwellers to photograph trees cut illegally with smartphones and upload the photos to link of the forest authorities. The forest link system is satellite based and therefore does not need mobile or internet connectivity. The volunteer forest watchers can upload the photos and make toll-free phone calls to authorities, the forest ministry, and the National Anti-Corruption Commission. The system has been an effective approach to curbing illegal logging in that, through this systems, the government of Cameroon collected over US $88,000 in fines for illegal forestry activities and revoked the licenses of four major logging companies in 2016. However, volunteer forest watchers are at high risk of being killed by the illegal loggers. According to the International human rights organization Global Witness, at least 185 volunteer forest, land, and river watchers were killed in 2015, which is the highest annual death toll recorded for people defending natural resources.

6.3.2 THE RAINFOREST CONNECTION (RFCX): CAMEROONIAN

Butler[7] writes that American-based non-profit company and London-based company (Zoological Society of London, ZSL) have installed the Rainforest Connection (RFCx) technology that can remit information in real time. This anti-deforestation technology was installed at several forests sites in Cameroon. The technology uses disused mobile phones, mounted on tree branches to sense and alert forest rangers on logging and poaching (Fig. 6.1). About 30 RFCx devices, recycled mobile phones monitor about 10,000 ha of forest land, covering an area of almost 40 square miles. The device listens out for sound connected to logging and poaching. The device is tuned to pick whir of a chainsaw, a gunfire, or sound of a moving truck. The audio signals are then transmitted to the command center so that the forest rangers can respond immediately. The several smartphones mounted on tree branches are powered by solar plates that are designed

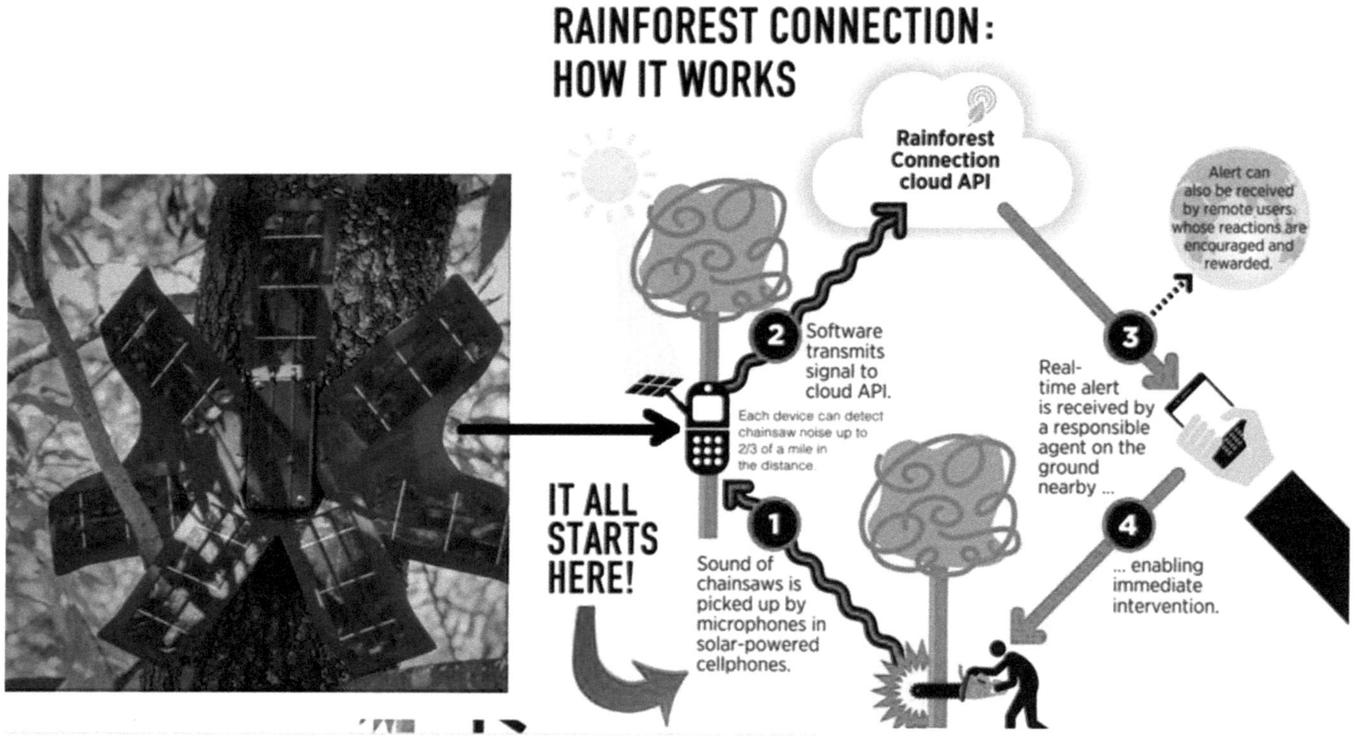

FIGURE 6.1 (See color insert.) RFCx device and its network connection

Source: Reprinted with permission from Ref. [6]. Creative Commons Attribution-NoDerivatives 4.0 International License.

for low-light conditions of the rainforest canopy. The units are durable so that they can survive hot and humid conditions for number of years. Each device is able to provide surveillance upto 300 ha of forest land (about 1 square mile).

Up to now, there are no published results of the project in Cameroon. However, the pilot scheme of the same project in Kalaweit forest reserve in Indonesia revealed that, within just 24 h of its deployment, the devices had already picked up sounds from illegal loggers and forest guards were immediately dispatched. Within 14 days of service in the forest, there was no illegal logging activity within the forest, neither did the loggers return after 1 year. However, just like any other technology, the device has some limitations, in that, powering the device has been a great challenge. Because of the densely covered tropical rain forest, direct light breaks through only in sporadic sun flecks. Typical solar photovoltaic arrays are compromised when even the smallest shadow falls upon them.

6.3.3 TIMBY: LIBERIA

TIMBY is a short form for "This Is My Back Yard." It is a real-time application that runs on a smartphone. It helps volunteer forest watches to report cases of illegal activities, and particularly, cases of forest clearing for oil palm plantation in Liberia. The application reports to the authorities, in real time, cases of illegal logging and other illegal forest activities.

Volunteer forest watchers can report illegal logging by uploading all forms of information, through videos, pictures, or audio messages on the mobile application. It, therefore, implies that volunteer forest watchers can prevent the illegal logging. The local forest watchers have already been trained on how best to use the technology. The program, in its pilot stages, is already showing promising results. For example, one of the companies that had been issued a license to carry out logging in Sinoe County had its license revoked because it was issued in error because the law was not followed. The team used TIMBY and reported the violation, which led to its revocation. Through TIMBY also, there have been more than 60 forestry concessions cancelled, millions of dollars of misspent county social development funds have been reported through the application, and it recently unearthed a $10.5 million scandal, where corporate funding did not go through government accounts.[15]

6.3.4 FOREST WATCH—UGANDA

According to Petersen and Pintea,[13] ForestWatcher is yet another ingeniously designed smartphone application that helps volunteer forest watchers monitor and report illegal logging in Kibale Forest in Uganda. The smartphone application is designed to allow easy, offline access to data about forest changes. Once the application is installed on a smartphone, it caches data to the phone, then the application directs users to a nearby forest clearing and enables the users to capture photos and fill out forms about deforestation location, which they can upload when back within internet reach. The application is unique in the sense that it displays forest change straight on the mobile devices in the hands of forest managers, indigenous communities, and law enforcement anywhere in the world, regardless of connectivity. ForestWatcher allows users to:

- Identify place(s) of interest to monitor.
- Download various satellite-based forest change and other contextual data (e.g., near real-time deforestation alerts, protected areas, and intact forest landscapes) onto a mobile device.
- Navigate to alerts in the field, even without the internet connection.
- Collect information including GPS points and photos through customizable forms.
- Review, analyze, and download data collected via the app.

Jane Goodall Institute (JGI) had already deployed an earlier version of ForestWatcher in Uganda since 2014. Hundreds of private forest owners, forest authority officers were trained on how to use mobile tools to navigate deforestation data in the field and collect evidence to support conservation action and this led to hundreds of prosecutions and fines for illegal forest loggers and encroachers.[3]

6.4 DISCUSSION

Despite the availability of other ways to fight illegal logging (such as lawsuits, government regulations, consumer boycotts, market pressure campaigns, and many others), high-tech surveillance is clearly one of the most effective ways to reduce illegal logging in Africa. There have been number of pilot projects across Africa and results from the pilot projects are quite promising. As already presented earlier, the forest cell phone monitoring systems piloted in Cameroon, Liberia, and Uganda are showing interesting results. For example, the ForestLink in Cameroon was able to curb illegal logging, in that, the government of Cameroon collected over US $88,000 in fines for illegal forestry activities and revoked the licenses of four major logging companies in 2016.[9] In Kalaweit forest reserve in Indonesia, the mobile phone devices had already picked up sounds from illegal loggers and forest guards were immediately dispatched, and 1 year later, illegal loggers had not returned.[6] Nuñez[12] confirms this success and observes that this technology has also been successfully piloted in Brazil, Ecuador, Peru, and Bolivia. The same promising results were seen in Uganda, whereby, through the ForestLink app, hundreds of prosecutions were carried out to collect huge sums of fines paid by illegal loggers and encroachers.

However, from the findings, there are also numbers of challenges of using mobile phone technology in combating illegal logging. Volunteer forest watchers who use the technology to report illegal logging become prime target to be killed by the illegal logger. The International human rights organization Global Witness observes that at least 185 volunteer forest, land, and river watchers were killed in 2015.[9] This, therefore, puts the lives of the volunteer forest watchers at risk. There are also challenges of deploying the technology itself. The densely covered tropical rainforest makes it difficult for the RFCx devices to get direct sunlight for the solar discs. Typical solar photovoltaic arrays are compromised when even the smallest shadow falls upon them. This has turned out to be a challenge.

6.5 SUMMARY

New wave of deforestation is currently sweeping across the entire continent of Africa destroying wildlife and incapacitating the resilience of its ecosystems to withstand the effects of climate change. There is still hope that this wave of terror aim at Africa's rainforest can be combated using mobile technology. Recent use of mobile phones to combat illegal forest logging has resulted in some interesting results in pilot projects in Cameroon, Liberia, and Uganda. Although substantial challenges still do exist, yet there are deliberate efforts to address the challenges.

KEYWORDS

- **Africa**
- **cell phone**
- **illegal logging**
- **timber**
- **tropical rain forest**

REFERENCES

1. *Africa Regional Review Report on Energy for Sustainable Development.* African Regional Implementation review for the 14th Session of the Commission on Sustainable Development (CSD-14); Prepared by the Economic Commission for Africa (UNECA) of behalf of the Joint Secretariat UNECA, UNEP, UNIDO, UNDP, ADB and NEPAD Secretariat, Rome, 2005; pp 1–50.
2. Aubréville, A. M. A. The Disappearance of the Tropical Forests of Africa. *Fire Ecol.* **2013,** *9* (2). DOI: 10.4996/fireecology.0902001 (accessed Jun 22, 2018).
3. Bhunia, P. New mobile app from World Resources Institute allows offline access to forest change data from Global Forest Watch. 205 Henderson Road #05-02A Singapore 159549; July, 2017, p 10. https://www.opengovasia.com/articles/8065-new-mobile-app-from-world-resources-institute-allows-offline-access-to-forest-change-data-from-global-forest-watch (accessed Jun 22, 2018).
4. Blaser, J. Forest Law Compliance and Governance in Tropical Countries, 2010. http://www.fao.org/docrep/012/al044e/al044e00.pdf (accessed Jun 22, 2018).
5. Boahene, K. The Challenge of Deforestation in Tropical Africa: Reflections on Its Principal Causes, Consequences and Solutions, June, 1998. https://doi.org/10.1002/(SICI)1099-145X(199805/06)9:3%3C247::AID-LDR278%3E3.0.CO;2-G (accessed Jun 22, 2018).
6. Butler, R.A. Discarded Cell Phones to Help Fight Rainforest Poachers, Loggers in Real-time. 2014. https://news.mongabay.com/2014/06/discarded-cell-phones-to-help-fight-rainforest-poachers-loggers-in-real-time/ (accessed Jun 22, 2018).
7. Cooney, D. Experts: New Wave of Deforestation Threatens Africa's Climate Resilience, 2011. http://www.fao.org/forestry/30071-0b6dab35a5dfc1cb9f4a628369 1f9e952.pdf (accessed Jun 22, 2018).
8. DFID (Department for International Development). Systematic Reviews in International Development, 2013. https://www.gov.uk/government/publications/systematic-reviews-in-international-development/systematic-reviews-in-international-development (accessed Jun 22, 2018).
9. Dunn, G. Smartphones Help Forest Defenders Fight Illegal Logging in Cameroon, 2017. http://www.borgenmagazine.com/fight-illegal-logging/ (accessed Jun 22, 2018).

10. Gulbrandsen, L. H; Humphreys, D. *International Initiatives to Address Tropical Timber Logging and Trade.* A Report for the Norwegian Ministry of the Environment, The Fridtj of Nansen Institute, Norway, 2006. https://www.files.ethz.ch/isn/47628/FNI-R0406_e.pdf (accessed Jun 22, 2018).
11. Minnemeyer, S. An Analysis of Access to Central Africa's Rainforests, 2002. http://pdf.wri.org/gfw_centralafrica_full.pdf (accessed Jun 22, 2018).
12. Nunez, C. Your Old Cell Phone Can Help Save the Rain Forest, May 2017. https://news.nationalgeographic.com/2017/06/topher-white-engineer-rainforests-explorer-festival/ (accessed Jun 22, 2018).
13. Petersen, R.; Pintea, L. Forest Watcher Brings Data Straight to Environmental Defenders, 2017. http://www.wri.org/blog/2017/09/forest-watcher-brings-data-straight-environmental-defenders (accessed Jun 22, 2018).
14. Ryan, P.; Tsunemine, K. Sub-Saharan Africa Report. Special Study Edition; Marubeni Research Institute, Marubeni, September 10, 2016, p 115. https://www.marubeni.com/en/research/report/subsaharan_report/data/24-Sub-SaharanAfricaSpecialEdition.pdf (accessed Jun 22, 2018).
15. The Goldman Environmental an App to Combat Illegal Logging, 2015. https://www.goldmanprize.org/blog/an-app-to-combat-illegal-logging/ (accessed Jun 22, 2018).
16. U.S. Forest Service. Illegal Logging and Forest Products Trade Undermine Economic, National, and Global Security, 2013. https://www.conservation.org/publications/Documents/CI_USG-Illegal-Logging-Case-Study.pdf (accessed Jun 22, 2018).
17. UCC (Uganda Communication Commission). Status of the Communications Market, www.ucc.co.ug, 2015.
18. UNCCD (The United Nations Convention to Combat Desertification). A Carrying Pillar in the Global Combat Against Land Degradation and Food Insecurity. Background chapter for the San Rossore meeting "Climate change: A New Global Vision," Pisa, Italy, July 15–16, 2004. http://www.sciepub.com/reference/19462 (accessed Jun 22, 2018).
19. UNESC (United Nations Economic and Social Council). Fifth Meeting of the Africa Committee on 2007. http://www.un.org/en/ecosoc/docs/newslet.shtml#year07 (accessed Jun 22, 2018).
20. World Bank Press Release. Regional Forest Law Enforcement and Governance (FLEG) Initiatives, 2013. http://www.worldbank.org/en/topic/forests/brief/fleg-regional-forest-law-enforcement-governance (accessed Jun 22, 2018).
21. Zanker, J. H.; Mallett, R. How to do a Rigorous, Evidence Focused Literature Review in International Development: A Guidance Note, 2013. https://www.odi.org/sites/odi.org.uk/files/odiassets/publications-opinion-files/8572.pdf (accessed Jun 22, 2018).

PART II

Selected Applications of Mobile Phones and Social Media in Education

CHAPTER 7

Operational Use of Mobile Learning: Understanding and Retention of Curriculum Content

JEREMY SUNKER, OYETADE KAYODE EMMANUEL, and KEHDINGA GEORGE FOMUNYAM

ABSTRACT

This chapter aims to evaluate and explore students' perceptions of mobile phones in helping with understanding and retention of curriculum content. This aim is achieved within a theoretical framework on the use of mobile learning with understanding and retention. Student learning is influenced by different environmental factors, which impact the adequacy of their understanding and retention of curriculum content. At tertiary education level, the intensity and quality of work become more physically and mentally demanding. Students may find it more difficult to deal with the workload they are presented with. As a result, this may affect the potential level of understanding that can be accomplished in certain subject areas. This study was conducted on undergraduate students studying Information Technology (IT) at a university in Durban, South Africa in 2016. The self-administered questionnaire-based survey was used while transactional distance theory (TDT) is the core theoretical framework that underpins this study. The outcome of students' perception was analyzed and it was found that the use of mobile phones in teaching and learning improved their understanding and retention of curriculum contents.

7.1 INTRODUCTION

An aspect to demonstrate mobile learning can be identified as using different features of the mobile phone such as Mp3, voice recordings, podcasting, videos, and pictures, which are easily accessible from the internet and can be downloaded directly onto any mobile device, provided the device has internet connectivity. This media can be viewed or listened at the convenience of the mobile user at any time and place. It is possible for mobile phones to be more easily integrated across the curriculum than desktops, and this is because many students already possess mobile phones.[14] The 2008 student survey by Evans[5] established that students consider learning materials in the form of podcasts on a mobile phone as an effective learning material and revision tool rather than traditional lectures, self-made study notes, and textbooks. It is interesting to note that mobile learning has the power to facilitate changes in the quality of learning modalities, which consequently can impact educational outcomes.[21]

Incidentally, m-learning is more than a mere extension of traditional methods of education. The m-learning aids alternative learning processes and instructional techniques that the concepts of new learning identify as effective for learning. It has been recognized that students face lot of challenges, such as time management, inadequate lecture time, and understanding of curriculum content.[21] This is intensified by their inability to balance studying with other day-to-day activities. More often than not, students find the workload at tertiary education level to be overwhelming. Finding the time to study while trying to complete assignments, homework, and personal work have an adverse effect on their studies. The operational use of mobile phones in tertiary education is one of the tools that can improve understanding and retention of curriculum content.

This study aims to uncover students' perceptions of mobile phones in helping to understand and retain the curriculum content. By understanding their perceptions, it provides a possibility to encourage the incorporation of mobile phones to assist students in improving understanding and retention at university level and to help deal with the abovementioned problems ultimately optimizing students' academic performance.

7.2 HYPOTHESES

Null hypothesis: Students perceive the operational use of mobile phones as a tool for learning to be ineffective in helping improve understanding and retention of curriculum content.

Directional hypothesis: Students perceive the operational use of mobile phones as a tool for learning to be effective in helping improve understanding and retention of curriculum content.

7.2.1 RATIONALE

This research was undertaken to find out students' perception of how the use of mobile phones can help with understanding and retention of curriculum content. If the directional hypotheses are correct, it will be possible to encourage the use of mobile phones based on the results of the research. The benefits of using mobile phones as a learning tool include "ease of use," which is the ability to use mobile phones to go over study materials anywhere at the student's own convenience.[5] For example, a student utilizing his or her time when traveling home on a bus to revise the curriculum content as well as the ability of using camera features of mobiles to take pictures of lecture notes during lectures instead of writing notes down from the board due to the high pace of lecturers. In addition, students are also able to use the voice note features on mobiles to record the voice of lecturers while they speak so they can go over it later and pick up important point they might have missed out during lectures.

It is believed that being able to repetitively go over revision work can improve retention of study material. Mobile phones are compact and comfortable to carry around. This has an advantage over using textbooks, which are much heavier and difficult to carry around.[4] It is easier to use a mobile device rather than textbooks, in terms of searching for pages and sections in the textbook for which one needs to study, whereas with podcasts or videos it is possible to just focus on a specific topic without having the trouble of going through a textbook.[16]

7.3 THEORETICAL FRAMEWORK

Several theories have been proposed in studying overall framework of the developments of technologies in distance education. These theories have focused on the use of technologies and their contributions to our understanding of distance education. Among these theories, this chapter focuses on the TDT as the core of framework that underpins distance education.

TDT is an educational theory that defines critical concepts of distance learning. It is a concept that describes the creation of teacher–learner relationships that exist when learners and instructors are separated by space and/or by time.[7,13] This interaction comes in the form of technology that helps to reinforce the transaction that fills up the communication space created by the distance.[13] TDT is "invaluable in guiding the complex practice of a rational process such as teaching and learning at a distance" (p. 3).[6] Hence, this theory is helpful in understanding the perceptions of students with the use of mobile phones in understanding and retention of curriculum content. TDT was chosen in this chapter because it informs about the material being studied by students. Using the theory to analyze their perceptions and how mobile phones can influence retention becomes an apt choice.

7.4 LITERATURE REVIEW

Asabere[1] argued that there is a growing need for mobile learning in Africa and the developing world in general. He articulates three reasons.

- Firstly, there is a disparity between existing academic facilities and physical infrastructure versus the increasing number of students admitted into educational institutions. This disparity creates the need or void which can be filled using mobile learning.
- Secondly, there is a limited time and place opportunities for workers. Other modes of education require students to be in the classroom or on site, which is most of the time difficult for employees as a result of their working schedules. This is supported by arguments by other scholars[22] that most institutions have limited staff and these staff members are already overburdened in ways that make it increasingly difficult for them to dedicate more time to other students or initiatives.
- Thirdly, some public institutions in developing countries are being originally developed as places for both study and residence because of their national character and model adopted. This leaves little room for more students as these campuses most often can only take few students.

The advancements in technology today have created an array of different means for teachers to interact with students.[17] Numerous tertiary institutions in South Africa have implemented the use of e-learning into their traditional teaching methods. The e-learning relates to the use of electronic media for a variety of learning purposes that range from add-on functions in conventional classrooms to full substitution for the face-to-face meetings by online encounters.[8] In essence, e-learning helps students with the freedom of pace, individual study, self-planned learning while the educator/institution provides guidance planning and feedback essential for continued student motivation, participation, and curriculum completion.[2,20] The e-learning in contrast to traditional lectures has the benefit of enabling learners to study at their own convenience (when, where, and how).

The use of m-learning naturally inherits the benefits of e-learning but extends its reach because mobile devices are portable and easily accessible to anybody and everybody.[5] m-Learning can be defined as "a form of e-learning, which can take place anytime, anywhere with the help of a mobile communication device such as a mobile phone, a personal digital assistant (PDA), iPod, or any such small portable device." This means that any portable digital or analogue device capable of keeping recordings and storing information can be used for m-learning. A key benefit of using a mobile device is that it allows students to study anywhere they prefer and to study "on the go.[9]" Other benefits are flexible and friendly environments leading to improved balance between students curriculum work and their lives; eliminating the demands of a scheduled study time and other commitments on and off campus reducing stress, anxiety, and absenteeism; a study tool that makes it easy for students to transport their study material anywhere; unanticipated free time as they regularly have their mobile handsets on them.[15]

In addition, learners are able to download study media directly to their devices so it is accessible at any given time and place providing greater flexibility to students which is highly beneficial in correspondence learning.[11] Students can also record lectures on their mobile devices and listen to them whenever they want instead of taking down notes. This is beneficial because when taking notes, it may be difficult to take note of curriculum content delivered by the lecturer. Furthermore, with the other features of m-learning such as podcasting, learning engagement is increased with better completion rates as students can watch videos downloaded to their mobiles and study materials repeatedly, which results in higher retention through revision.[12,19] With podcasting, students find it more fun to use their mobile devices to study, it replaces the boring methods of having to sit with books. Other benefits are multidevice support because audio and video media are compatible with almost all mobile devices today. Some of the challenges faced with podcasting are lack of quality recording as lecturers may be recording the podcasts in an environment that is not acoustically suitable.[18]

Investigators[1,22] argue that there are numerous benefits of using mobile learning in higher education (HE), such as: (1) Mobile learning can occur at anyplace and anytime, and learning content can be accessed anywhere. This makes mobile learning unique and efficient; (2) mobile learning processes are not limited to one particular place. The experience of learning can happen anywhere; (3) mobile learning enhances interaction between instructors and learners/students. Students are able to interact with the people facilitating the curriculum any time they are studying and need help; (4) mobile learning is a great opportunity for students and instructors to continue to learn while on the move. Just like the name, students can go anywhere with the learning experience; and (5) students in mobile learning practice undergo self-centered learning with focus. Self-centered or learner-centered learning give students the opportunity of pacing their own learning and making the best out of every opportunity they get.

Furthermore, mobile learning is a great opportunity for just in time training or review of content. The content can be reviewed and reconceptualized at any time to meet the students changing needs. Finally, mobile learning facilitates collaboration among students and instructors through both asynchronous and synchronous communication techniques. The synchronization between the devices ensures that all stake holders get feedback at the same time.

In conclusion, learning in higher education is seen as an individual effort. Most students vary greatly in their learning styles (abilities and disabilities) and assimilate at different rates. It is therefore important to assist these students with the needed tools that would allow them to learn at their varying pace and timeframes.[2,10]

7.4.1 CURRENT STATUS OF RESEARCH

Technology is constantly changing and evolving at a fast rate, there are new ways or forms of media that can be used for learning such as virtual reality. There have been advancements in terms of quality and availability of videos and podcasting. Evidence from literature shows that most studies conducted relating to using mobile devices as a tool for education are done in the developed world.[3,17] As countries have different standards of education, it is important to conduct this research in the South African context so that readers may be able to understand or identify the problem or opportunities for improvement with regards to education.

7.5 RESEARCH METHODOLOGY

Research strategy employed in this study used the quantitative research approach and the method for gathering data was a questionnaire. The total population size of 70 Bachelor of Technology (BTech) and IT students from the Faculty of Accounting and Informatics were targeted for the survey. Applying this sample size to the formula below[17] gives the calculation of our sample size proportion.

$$n = \frac{NZ^2 P(1-P)}{d^2 (N-1) + Z^2 P(1-P)} \tag{7.1}$$

where, $N = 70$ (total population), $Z = 1.96$ (statistic for level of confidence), $P = 0.05$ (expected prevalence or proportion), $d = 0.05$ (expected level of precision), sample size $(n) = 35$.

Accidental nonprobability sampling method was used. The implementation of the accidental nonprobability sampling method involved using random volunteers from the BTech class. A total of 35 survey responses were obtained. Survey results were analyzed using descriptive statistics and frequency response (mode or percentage of responses). The aim of using frequency response was to observe where there are areas of strong and weak correspondence among the results obtained.

7.6 RESULTS

The surveyed IT students evaluated the perception of students on the use of mobile phones for educational purposes and use of mobile phones in understanding and retention of curriculum content via an anonymous paper-and-pencil survey. From the survey, insight was gained regarding the students' perception of the effectiveness of mobile phones as a helpful tool for effective learning.

7.6.1 DATA RELIABILITY AND VALIDITY

The Cronbach alpha method was used to provide the measure of consistency scale within the questionnaire. Results in Table 7.1 show that data collected from students who participated in this study are reliable (Cronbach's alpha (α) greater than 0.75).

TABLE 7.1 Reliability Table for Research Variable.

Research variable	No. of items	Cronbach's alpha (α)
Students' perceptions of mobile use for educational purposes	10	0.95
Students' perceptions of mobile use in understanding and retention of curriculum content	7	0.77

7.6.2 DESCRIPTIVE STATISTICS

Table 7.2 gives an overview of the students' demographics. The study found that a simple majority of the respondents of this survey are males (60%) between ages 20 and 23 (79%). There is almost a mix of ethnicity between students of Indian and African origin who strongly use mobile phones on a daily basis (97%). Table 7.3 represents the perceived usefulness of mobile phones for educational purposes by students who participated in this study.

Results in Table 7.3 indicated that students overwhelmingly agree (79%) that the use of mobile phones for educational purpose is effective and useful. Descriptive statistics for individual questions were analyzed nonparametrically using the mean. In response to the statement "I think that using mobile phones as an educational tool can enhance my learning in general," the majority of students (85%) agree that mobile phone was useful. Out of the 35 students, 74% of them indicated that it was important for them to be able to access curriculum content through the use of their mobile phones and 65% of respondents indicated that using their mobile phones to acquire their study material was necessary. This implies that in general, the majority of students in this survey overwhelmingly agree (84%) that mobile phones are useful tools for educational purposes.

TABLE 7.2 Profile of the Respondents.

Demographic characteristic	Respondents, %	
Gender	Male	60
	Female	40
Age groups	17–19	0
	20–23	79
	24–26	13
	Above 27	8
Mobile usage frequency	None	0
	Daily	97
	Weekly	3
	Monthly	0
Ethnicity	African	44
	Indian	50
	Colored	0
	White	0
	Other	6

TABLE 7.3 Students' Perceptions of Mobile Use for Educational Purposes.

B	SA	FA	WA	FD	SD	Std Dev	Skewness
B1	62	17	6	6	9	8.7	1.99
B2	43	15	11	17	14	4.6	2.19
B3	49	11	14	17	9	5.76	2.12
B4	46	23	11	9	11	5.59	1.5
B5	51	20	11	15	3	6.61	1.7
B6	40	26	6	17	11	4.76	0.95
B7	43	14	17	17	9	4.71	1.93
B8	51	9	14	17	9	6.34	2.1
B9	65	17	6	9	3	9.26	2.01
B10	71	20	0	6	3	10.54	1.89
Ave	52	17	10	13	8	–	–

Table 7.4 represents students' perceptions of mobile phone use in understanding and retention of curriculum content. The results from this table indicated that the students surveyed overwhelmingly agree (84%) that the use of mobile phones is an effective tool that assists with the understanding and retention of curriculum contents. Descriptive statistics for individual questions were analyzed nonparametrically using the mean. Results show that 86% of students surveyed agreed that watching YouTube tutorials of curriculum content downloaded on their phones was an effective means of helping them with curriculum content they did not understand. Also, 83% of respondents agreed that watching tutorials of curriculum contents on their mobile phones helps them reinforce learning content they have learnt in class.

TABLE 7.4 Students' Perceptions of Mobile Use in Understanding and Retention of Curriculum Content.

C	SA	FA	WA	FD	SD	Std Dev	Skewness
C1	49	23	11	9	9	6.14	1.58
C2	51	17	9	20	3	6.61	1.66
C3	57	23	6	9	6	7.82	1.71
C4	49	23	11	9	11	6.38	1.21
C5	43	29	14	9	6	5.63	0.833
C6	49	31	9	9	3	6.94	0.96
C7	49	14	26	9	3	6.5	1.56
Ave	50	22	12	10	6	–	–

Overall, the majority of surveyed students (84%) agree that mobile phones are useful tools that assist with understanding and retention of curriculum content. These results tested the hypothesis by comparing how they perceive it as a tool for education and how they perceive it in assisting with understanding and retention which show a higher positive deviation in the responses confirming our directional hypothesis. Significantly, most students believed that mobile phones are a useful tool in tertiary education purposes and it does assist in understanding and retention.

7.7 DISCUSSIONS

This study builds on the knowledge that majority of students find academic workload in University very demanding. Hence, this chapter intended to examine the perception of students on the operational use of mobile learning in helping with understanding and retention of curriculum content. The results in Section 7.6 have provided a general overview of the perceptual experience of students in this respect.

In a nutshell, the results of this study can be presented as follows:

- According to the literature reviewed in this study, transactional distance theory is able to explain the perceptions of students on the use of mobile learning in helping with understanding and retention of curriculum content.
- According to the results of the survey conducted in this study, students perceived the use of mobile phones is effective and useful for educational purpose.
- According to the results of the survey conducted in this study, students overwhelmingly agree that mobile phones assist with understanding and retention of curriculum content. This satisfies the aim of this study and confirms the directional hypothesis.
- There seems to be a general agreement between the findings of this study and existing literature on the use of technologies in education. These technologies reinforce or enhance motivation, including self-direction. It also offers a more friendly and supportive atmosphere than the traditional teaching.[14,16]
- This study found that the number of female IT students is very low compared with their male counterparts. Therefore, measures need to be put in place to encourage active participation of female students in academic-related activities.

7.8 SUMMARY

South Africa is a developing country and is faced with challenges that are similar to other developing nations with regards to education. Not all students have personal access to personal computers or laptops, and so using their mobile phones becomes an alternative. Taking advantage of this useful tool can assist in helping students with their academic work. According to the results uncovered in this chapter, students perceive mobile phones not only as a useful tool for education but also as a means to use it for helping improve understanding and retention. Boundaries of technology are continually expanding and advancing at an exponential rate. With this improvement in mobile technology, it opens endless possibilities of enhancing education in South Africa. Exploiting this mobile technology would be a next step in uncovering the possibilities of improvement in education.

KEYWORDS

- **e-learning**
- **m-learning**
- **mobile phone**
- **MP3**
- **podcast**
- **South Africa**

REFERENCES

1. Asabere, N. Y. Benefits and Challenges of Mobile Learning Implementation: Story of Developing Nations. *Int. J. Comput. Appl.* **2013,** *73* (1), 1–10.
2. Badu-Nyarko, S. K. Isolation and Control in Distance Education: The Case of the Ghanaian Student. *Int. J. Instruct. Technol. Dist. Educ.* **2010,** *7* (3), 13–24.
3. Deal, A. Teaching with Technology White Paper, Podcasting. *Teaching Technol.* **2007,** *6,* 1–15.
4. Deal, A. Podcasting. *Internal Reports and Documents*; Carnegie Mellon University Press: Pittsburgh, USA, 2007; p 18.
5. Evans, C. The Effectiveness of m-learning in the Form of Podcast Revision Lectures in Higher Education. *Comput. Educ.* **2008,** *50* (2), 491–498.
6. Garrison, R. Theoretical Challenges for Distance Education in the 21st Century: A Shift from Structural to Transactional Issues. *Int. Revi. Res. Open Distrib. Learn.* **2000,** *1* (1), 1–10.
7. Gorsky, P.; Caspi, A. Critical Analysis of Transactional Distance Theory. *Q. Rev. Distance Educ.* **2005,** *6* (1), 1–11.

8. Guri-Rosenblit, S. Distance Education and e-learning: Not the Same Thing. *High. Educ.* **2005,** *49* (4), 467–493.
9. Kadirire, J.; Guy, R. Mobile Learning Demystified. In *The Evolution of Mobile Teaching and Learning*; Informing Science Press: Santa Rosa, CA, 2009; pp 15–56.
10. Knowles, M. S., Ed. *The Modern Practice of Adult Education.* New York Association Press: New York, 1970; p 400.
11. McGarr, O. Review of Podcasting in Higher Education: Its Influence on the Traditional Lecture. *Australas. J. Educ. Technol.* **2009,** *25* (3), 309–321.
12. Mehdipour, Y.; Zerehkafi, H. Mobile Learning for Education: Benefits and Challenges. *Int. J. Computat. Eng. Res.* **2013,** *3* (6), 93–101.
13. Moore, M. G. Theory of Transactional Distance. *Theoret. Princ. Distance Educ.* **1993,** 1, 22–38.
14. Moseley, D.; Higgins, S.; Bramald, R..; Hardman, F.; Miller, J.; Mroz, M.; Williamson, J. *Ways Forward with ICT: Effective Pedagogy Using Information and Communications Technology for Literacy and Numeracy in Primary Schools.* University of Newcastle Press: Newcastle, United Kingdom, 1999; p 130.
15. Nordin, N. M.; Hamzah, M. I.; Yunus, M. M.; Embi, M. A. The Mobile Learning Environment for the In-service School Administrators. *Procedia Soc. Behav. Sci.* **2010,** *7*, 671–679.
16. O'Bannon, B. W.; Lubke, J. K.; Beard, J. L.; Britt, V. G. Using Podcasts to Replace Lecture: Effects on Student Achievement. *Comput. Educ.* **2011,** *57* (3), 1885–1892.
17. Oyetade, K. E.; Eyono Obono, S. D. Perceptions of Educators on the Learners' Youngest Age for the Introduction of ICTs in Schools: Personality Theory Approach. *Int. J. Soc. Behav. Educ. Econ. Business Ind. Eng. (World Acad. Sci. Eng. Technol.)*, **2014,** *8* (11), 3574–3580.
18. Park, Y. Pedagogical Framework for Mobile Learning: Categorizing Educational Applications of Mobile Technologies into Four Types. *Int. Rev. Res. Open Distrib. Learn.* **2011,** *12* (2), 78—102.
19. Resell-Aguilar, F. Podcasting as a Language Teaching and Learning Tool. In *Ten Years of the LLAS e-learning Symposium: Case Studies in Good Practice*; Borthwick, K; Corradini, E.; Dickens, A., Eds.; Research-publishing.net: Dublin, Ireland, 2015; pp 31–39.
20. Sewart, D. One World, Many Voices: Quality in Open and Distance Learning. In *17th World Conference of the International Council for Distance Education*; International Council for Distance Education Birmingham: United Kingdom, 1995; p 516.
21. Valk, J. H.; Rashid, A. T.; Elder, L. Using Mobile Phones to Improve Educational Outcomes: An Analysis of Evidence from Asia. *Int. Rev. Res. Open Distrib. Learn.* **2010,** *11* (1), 117–140.
22. Wang, M.; Ng, J. W. Intelligent Mobile Cloud Education: Smart Anytime-Anywhere Learning for the Next Generation Campus Environment. In *8th International Conference on Intelligent Environments (IE)*; June 2012; pp 149–156.

CHAPTER 8

WhatsApp-Based E-Learning in Adult Education for Lesotho's Inmates: Methods and Evaluation

SELLOANE PITIKOE

ABSTRACT

The 21st-century world has become a globalized digital village, where mobile device applications such as WhatsApp messenger have transformed traditional classroom learning into an epistemic digital learning mode. Notably, WhatsApp messenger is applauded for transforming traditional classroom learning into an epistemic digital e-learning that widens access to education beyond the confines of time and space. This chapter addresses two research objectives: (1) it investigates how Lesotho's rehabilitation education for incarcerated inmates is offered and (2) it examines the effects of integrating WhatsApp messenger as a learning tool in different forms of education including inmates' education programs. The author used literature review approach to examine the effects of WhatsApp messenger as a digital learning tool. WhatsApp messenger is gaining popularity both as mobile social media application and an e-learning platform since its inception in 2010. Nonetheless, the paucity in literature on integration of WhatsApp messenger in Correctional Services Education Programs still remains. The potential challenges in integrating the application as a tool for facilitating e-learning are also underlined. The chapter concludes by recommending areas for future research studies.

8.1 INTRODUCTION

The world is flooded by massive use of mobile devices both for socialization and conducive flexible learning environment where collaboration, interaction, immediate communication, information, and resource sharing become pivotal.[6–8,10,13,15,16,26,27,30,31,33,37,39,43,41,46] Mobile digital devices such as smartphones, tablets, notebooks, and iPads can be incorporated in all different forms of education provision.[7] They are portable and come with communication applications which also open new opportunities for learners to develop social skills such as video conferencing, video calling, and chat groups.[22]

Information Communication Technology (ICT) is an umbrella word denoting all technological forms (hardware and software), which capture, process, store, and digitally transmit information.[13] E-learning is defined as ICT-assisted learning provided through either wireless or mobile devices such as virtual learning platforms. There are three categories of e-learning, namely, distance learning; software-assisted teaching and learning, and electronically mediated learning. These tools are reportedly user-friendly, economic on time and money, and have access restrictions.[22,41]

Scholars define social media differently. For the purposes of this chapter, Ref. [16] define social media as internet applications grouped together against the backdrop of Web 2.0 to facilitate digital creation and content exchange through well-established digital networks and devices. In other words, technology becomes resourceful in promoting an interactive learner-centered teaching and learning practice. There are five key components of social media, namely:[16] (1) social networking sites such as Facebook and LinkedIn; (2) media sharing sites like YouTube; (3) creation and publishing tools such as Wikis; (4) aggregation and republishing through RSS tools; and (5) remixing and republishing tools. Using social media technologies improves learner performance through peer learning and keeps the learners abreast with the latest developments in technology.

Regardless of the mounting recognition of the flexibility and affordability of technology-based learning tools and their usefulness in facilitating collaborative learning, the appreciation of social media in the correctional service education programs remains invisible. Likewise, the benefits of e-learning are yet to be realized by the correctional facilities in developing countries, including Lesotho. Research also indicates that inmates possess mobile devices upon entry to correctional services.[21] However, the discourses of bureaucracy and policy hegemony demand an impounding of their personal belongings including mobile devices "[cutting their] access to 'smart' phones, tablets, or other internet enabled mobile devices" for safety and security until completion of the sentence.

Literature also unveils that incarcerated inmates are generally happy with the rehabilitation programs that they receive. However, their limited access to technology coupled with security routines tend to impede their full access and participation to rehabilitation programs.[11] Nonetheless, scientific research on the reasons preventing incarcerated inmates' access to technology-based learning is very thin on the ground. On this background, an intensive review of literature was conducted in order to examine the effectiveness of WhatsApp messenger as a learning tool that can be integrated within the rehabilitation programs offered by Lesotho's Correctional Services (LCS).

This chapter explores the potential of using WhatsApp messenger as a digital tool for facilitating e-learning-based rehabilitation programs offered by LCS.

8.2 BACKGROUND

Lesotho is one of the smallest kingdoms located in Southern Africa and a former British protectorate that gained independence on October 4, 1966. The total area of Lesotho is 30,355 km^2 while the population is about 2 million. Lesotho is located at the center of South Africa creating an intimate connection between the two countries' economy; in that, the slightest economic discomfort experienced by her only neighbor, South Africa affects Lesotho. South Africa exports 80% of its commodities to Lesotho while Lesotho only exports a quarter of her commodities to South Africa. This unhealthy economy positions Lesotho as a labor warehouse for South Africa's mining and industry while the latter provides skilled scientific researchers as consultants in Lesotho when the need arises.

The policy environment reveals how the Goal 4 of the sustainable development goals (SDGs) highlights a need for equitable quality and inclusive lifelong learning and sustainable livelihoods. However, the latest SDGs progress report indicates a significant 9% stagnation dating back to 2008 and remarkably high levels of exclusion and hard-to-reach learner population. This notion closely resonates with incarcerated inmates; in that, their institutional policies and regulations restrict their accessibility to the external world beyond the correctional service confines. The international policy background compelled a global partaking in the development of customized ICT policies that would open doors to a flexible universal education provision independent of space and time constraints for ease of access by all groups of society including inmates. In response to the global outcry, the government of Lesotho launched Lesotho's first ICT policy in 2005 with a vision to broaden access to ICT by all members of society. Nonetheless, incarcerated inmates continue to be denied access to mobile devices by the prison bureaucracy in the interim, not only hampering their interaction with the outside world but also limiting their participation in digital learning communities.

Education provision in Lesotho is offered through two main modes of education: formal and non-formal (NFE). Formal education is offered through public and church-administered schools where the government functions as the overall administrator of the national education provision. NFE on the other hand is offered mainly by the non-governmental organizations (NGOs) in Lesotho and receives moderate government support as it is regarded to be of inferior status as opposed to formal provision. NFE is a flexible form that widens access to the marginalized underserved social groups and inmates. The integration of technology can further bring education closer to the learners whose predicaments prevent normal access such as inmates.

Notably, correctional service in other countries as well as in Lesotho aims to capacitate the inmates with the necessary life and production skills that would help them to make better choices in future, be employable, and avoid reoffending.[23,25] Well-structured, managed, and coordinated rehabilitation programs benefit both the individual inmates and the nation at large. For instance, while inmates acquire transformative skills, in the long term it reduces the social costs of crime, while also supporting rehabilitation of inmates for their community integration and re-entry.[6,20,28,42,44] Some prisoners hate the idle time that they spent during incarceration and want to put it into good use by continuing to study while serving.[42]

Therefore, this calls for an accommodative, high quality provision that is aligned with the millennial education standards and tailored to meet both the social and economic needs of the inmates.[6,20,42] Rehabilitation programs can come in different forms in order to address the diverse learner needs. For example,[5,29] it is argued to include music in rehabilitation programs to cater for learning and fun while also promoting talent search among the inmates.

Notably, LCS provides NFE rehabilitation programs for inmates[28,33] in partnership with the local civil society organizations (CSOs) and NGOs. In some cases, the rehabilitation programs are also offered by LCS staff and inmates.[33] Some of the challenges facing the inmates' rehabilitation programs in Lesotho include limited resources (material, financial, and human capital) and a mismatch between sentencing and the rehabilitation program schedule.[28] For instance, short-term sentenced inmates are scheduled in the long-term programs; hence, they complete their sentence before completing the program. This challenge presents a need for rehabilitation programs that continue beyond the sentencing timeframe to prepare the inmates with market related skills to cope outside the prison walls. Scholars lament that the LCS-offered NFE provision is less concerned about human development and the benefit that come with it. Instead, it is used as a social control mechanism aimed to distract the inmates from idling.[33] In other words, the provision is more of an institutional punitive control measure instead of providing a habitable and rehabilitative environment that enables inmates to safely re-enter into their society upon completion of their sentences.[33] Additionally, the facilities suffer from poor or lack of management which threatens the sustainability of such rehabilitation programs.

Nonetheless, inmates like the rest of the society are entitled to access millennially befitting high quality, equitable, and holistic education as a basic human right enjoyable by all humans.[10,20,28,38,39,42,45] Therefore, given the dynamics of their confinement, the author advocates for integration of e-learning mobile applications such as WhatsApp messenger. This digital tool can facilitate learning continuity for the ex-convicts while also giving them the freedom to learn and share within a collaborative digital community without confining the inmates to a specific space and time.

8.3 LITERATURE REVIEW

8.3.1 WHATSAPP MESSENGER

The section signifies e-learning devices and their role in transforming the traditional classroom confined teaching and learning practice into an autonomous learning space that are accessible anywhere and anytime. The portability of the mobile devices and the applications that they come with facilitate ease of teacher–learner communication and a collaborative access to information and learning resources. The e-learning platform argued for in this chapter is the WhatsApp Messenger. WhatsApp is an English origin phrase "What's up? [or] What's new?"[4,5]

WhatsApp is closely linked with the user's smartphone which means that, in order for one to access the messenger, one has to create an account using a smartphone. Immediately after creating the WhatsApp account, the user can access free instant messages which make the application even more attractive and affordable.[3,15,26,37,46] The application has recently strengthened user privacy by introducing the end-to-end encryption which ensures that the user's content is protected.[8]

WhatsApp tool widens access to interactive learning beyond the classroom confines in a bid to address the needs of the underserved social groups such as inmates. Effective use of the WhatsApp messenger requires internet data usage which can either be from the user's smartphone or it can be browsed from desktops and laptops.[8] The digital information sharing becomes the sole responsibility of the learning community members to enable collaborative access to learning resources between the learners and the teachers, irrespective of distance and time.[37]

This instant messaging application was developed in 2009 by Brian Acton and Jan Koum[46] as an upgraded form of the short messaging system (SMS) that was faster and easier digital information sharing tool using multimedia modes both for personal and academic social learning.[8,37,46] In 2016, WhatsApp was absorbed within the Facebook Inc. family. This messenger comes with easy-to-use applications such as group chat, audio notes, sending and receiving messages, as well as voice and video calling.[3,4,15,26,37,46] Notably, the group chat allows users to send up to "256 instant messages."[4,8] While it is argued that WhatsApp is still at an infancy stage as a social media technology; the application has attracted a lot of scientific research interest from various scientific fields of study on its potential role as an e-learning tool.

8.3.2 USES OF WHATSAPP MESSENGER IN LEARNING

Studies revealed compelling evidence where WhatsApp messenger was used as a tool for learning in different scientific fields. Generally, the literature points to the affordability of WhatsApp in that the instant messages can be accessed freely. Literature also highlights that the messenger is easy to use and promotes self-directed independent and interactive learning accessible at the learner's convenience and time. WhatsApp messenger enables learners to acquire soft skills such as communication through chat groups, video and audio messaging facilities. The web browsing search programs allow for acquisition of technical skills, which are critical in the labor market.

Research that investigated the effectiveness of WhatsApp messenger as a learning tool highlight positive effects of WhatsApp messenger in learning. Researchers, who investigated the effects of WhatsApp in English language learning, found an improved performance, confidence, and motivation by learners.[3,37] The application was also applauded for its ability to support independent language learning and freedom of expression outside the controlled classroom environment.[4,15] The usefulness of WhatsApp in studying English as second official language (TESOL) revealed that students preferred WhatsApp messenger because it is affordable, supports interactive learning, enhances learners' technological skills, and how text messaging was the most favored peer communication method among learners.[4,15] WhatsApp mainly has four functions: learner motivation, social networking, digital information sharing, and digital time planning management.[4] Similarly, WhatsApp application on Arabic language learners discovered that integration of WhatsApp messenger enabled a three-way collaboration: firstly, among students and their peers; secondly, between students and content; and thirdly, between students and teachers.[1]

The clinical environment of the United Kingdom on how WhatsApp not only opened access to effective communication among the health professionals[30] but also the application opened a collaborative learning platform for students and teaching staff with opportunities to seek clarification, prepare for tests using the search mode, and download the course modules at their own convenient time. Also, the recently introduced end–end encryption was also a resourceful confidential tool to provide privacy while also protecting the content.[30]

These findings highlight the positive effects of WhatsApp in learning, such as learner motivation, confidence enhancement, support to independent learning, allows for unrestricted freedom of expression, affordability, collaborative learning, technological skills acquisition, and the ability to use text messaging for peer communication. Generally, the key functions of WhatsApp are learner motivation, social networking, digital information sharing, and digital time planning management.[4]

While these findings were mostly from the Western countries, yet they can be potentially adapted to suit the African context. Lesotho demonstrated the feasibility of integrating WhatsApp in various learning fields.

This characteristic can arguably meet the requirements of the rehabilitation education provision provided by LCS facilities, which form the main argument of this chapter.

Other scholars, however, lament that the integration of WhatsApp messenger also has negative effects on the learning. Literature revealed that online learning distracts learners.[1,46] Participants lamented that more time was consumed on chatting and socializing with friends, families, and significant others.[46] Additionally, WhatsApp messenger learners complained about the alarming tendency of users to shorten words when messaging, spelling, grammatical errors, and poor sentence construction. Also, the time spent on online social networking consumed valuable time that could be committed to assignments completion.[46] The application was also reportedly expensive and bringing in an additional study workload.[1] WhatsApp can also be addictive to its users[2] because users spend too much time online which if uncontrolled can easily become an uncontrollable addiction.

In spite of the limitations, WhatsApp messenger comes with more advantages that outweigh the challenges. The fact that the application is closely linked to mobile devices allows for free instant messaging: a feature that makes it more suitable to effectively meet the needs of the underserved members of society such as the incarcerated inmates of Lesotho. Also, its portable size can enable learning all the time regardless of the rugged topography of Lesotho, which largely limits access to social services including education. On this background, the usefulness of WhatsApp messenger for rehabilitation programs can be extended beyond the correctional service sentence for continuity and sustainability.

Scholars identify four issues for consideration in regard to implementing social media in adult education programs:[1,2,4,15,46]

- Firstly, the educator must be well versed in digital skills in order to develop content that is appealing and attractive to the learners.
- Secondly, the learners also have to be digitally competent in order to be able to access the material and apply the contents to learning.
- Thirdly, the chosen platform must allow learner-to-learner interaction as well as learner–educator interaction.
- Fourthly, the quantity of social media content has to be less than the classroom material to allow ease of sharing and discussions by the stakeholders.

8.4 MATERIALS AND METHODS

The discussions and arguments presented in this chapter support the proposed e-learning-based education provision that integrates WhatsApp messenger in learning programs intended for inmates in Lesotho. This study followed a systematic literature review for its ability to synthesize qualitative evidence.[31,35] Selective literature was engaged[31] to achieve the research objectives stated such as:

- First, it investigates how Lesotho's rehabilitation education for incarcerated inmates is offered.
- Secondly, it examines the effects of integrating WhatsApp messenger as a learning tool in different forms of education including inmates' education programs.

The findings harvested from this chapter can help Lesotho and other countries with similar study interest to transform rehabilitation education provision intended for inmates into an interactive and relevant to their learning needs by integrating digital tools. Another data set was harvested from journal articles which presented the integration of WhatsApp messenger, an instant messaging e-learning tool. This set sheds more light on the effects of WhatsApp messenger and the potential challenges that may hinder its integration in rehabilitation education for inmates. Since the chapter is set against the backdrop of prison environment, it was crucial to review literature as an overview of the current rehabilitation education provision by correctional services. The literature searches were on studies related to social media, media devices, prison education, as well as WhatsApp and their use in all forms of teaching and learning including prison rehabilitation programs.

8.5 RESULTS AND DISCUSSION

8.5.1 HOW LESOTHO REHABILITATES INMATES: OBJECTIVE 1

A global overview regarding provision of rehabilitation programs for incarcerated inmates revealed that correctional services offer their programs using the traditional classroom-based teaching and learning practice. It was also discovered that the rehabilitation programs are aimed to capacitate the inmates with vocational skills that prepare them for the world of work upon completion of their sentence while also preventing them from recidivism. The success of these programs is reportedly hampered by the red tape and the frequent correctional services' security checks which distract the inmates' concentration.

Rehabilitation programs offered in Lesotho come through the partnership that LCS has with the CSOs and NGOs working in Lesotho. Poor management of the provision coupled with limited access to resources has rendered the provision unfocussed so much that it has been relegated to a bureaucratic social control mechanism that prevents the inmates from idling. It was also unveiled that completion and continuity of the offered rehabilitation programs in Lesotho becomes a problem in that there is a mismatch between the inmates' sentencing and the scheduling of the rehabilitation programs. Those inmates who are released before completing the course have limited access to the course content outside the institution in the process which threatens their ability to explore the job market. In the end, poverty and unemployment compel them to reoffend.

The literature indicates that the inmates want to commit their idle time into learning some skills that can be applicable upon their community re-entry and socialization process. However, the straight jacket nature of correctional service governance prohibits them access to quality education that meets the requirement of the millennial community where integration of technology widens access to social networking tools and platforms for collaborative learning. This social disconnect shuts them off from the rest of the different forms of communities and their opportunities to engage in learner-centered capacity building activities of their choice. Literature on correctional services rehabilitation programs offered in Lesotho is very scarce to inform education policy and practice on the current global trends in education that Lesotho can consider in order to make the rehabilitation programs become more attractive and appealing to the inmates.

The chapter reveals that inmates are constitutionally entitled to enjoy the basic human rights by virtue of being human including quality education[17,18] and the education provision has to be designed with the learner being central to the design process.[10,20] Given the meagre literature availability challenges, Lesotho academia members from different scientific fields to engage in scientific research discussions challenging the status quo regarding provision of rehabilitation programs in Lesotho into more of accommodative learning provision that is of good quality, widely accessible, cost-effective, learner-directed, available anytime, and anywhere to ease access to marginalized and underserved society members such as inmates.

8.5.2 E-LEARNING THROUGH WHATSAPP MESSENGER: OBJECTIVE 2

Generally, the literature findings point to the benefits of WhatsApp as an e-learning tool through its three main functions:[1]

- First, it facilitates peer to peer interaction, where students can independently learn from their peers at their own time and space in the absence of the teacher.[1] This approach allows them freedom to unrestricted interaction without the classroom setting. It also opens a window of opportunity for introverted learners to express themselves within a friendly and conducive peer learning environment. Additionally, teachers can collaborate with their peers either internally within the same institution, module, cohort or externally with other like-minded colleagues for digital content creation, sharing, and new ideas.
- Secondly, WhatsApp allows students to interact with the content in collaboration with other members of the e-learning groups.[1] WhatsApp comes with facilities such as audio notes, video messaging, camera, and instant messaging, which can be used to share digital information among the leaning community,[2,7] whereby learning becomes a community activity and a common good for all where members are expected to make a significant contribution in the form of digital information and resource sharing. The web browsers can provide the learners with an opportunity to search relevant digital content and share with their colleagues. It also gives the learning community members a copyright of the shared information.
- Thirdly, the digital platforms allow for interaction between teachers and students through platforms such as chat groups where teachers can upload assignments for students to complete, post pertinent instructions for learners,[2] post assignment feedback on the completed assignment, module content, scheduling of events and updates. In turn, students may also use these platforms to download assignments, modules, and ask clarity seeking questions.

In as much as WhatsApp messenger can potentially benefit Lesotho's inmates during and after sentencing, care has to be taken that the current civil unrest and political instability can threaten the successful integration of WhatsApp in Lesotho's rehabilitation programs. Another glaring challenge is the consistently declining national economy, which becomes an impediment to resource allocation to support the integration of e-learning platforms by LCS. While ICT can potentially increase access to learning for Lesotho's inmates, unlimited access to ICT literacy remains pertinent.

Some of the gaps in the literature findings include the current limited evidence-based epistemological research on WhatsApp use in correctional service context to inform policy and practice agenda change with regards to rehabilitation education provision in other countries as well as in Lesotho. Additionally, the limited knowledge generated on rehabilitation programs in Lesotho is also underscored. These gaps bear heavily on remodeling Lesotho's rehabilitation programs and curriculum reform in such a way that they not only empower the inmates while incarcerated but also continue to be resourceful in their future endeavors beyond prison. However, linking social capital that LCS has established with the CSOs and NGOs can become useful and cost-effective resources to tap; and explore the feasibility of integrating WhatsApp within the rehabilitation programs. On this backdrop, the author suggests further investigation on the inmates' preferences of media devices for e-learning with regard to choice, accessibility, as well as cost in order to inform the curriculum transformation. Another gap that needs further interrogation through scientific research concerns the unforeseen potential barriers that inmates may face when it comes to e-learning in Lesotho.

8.6 LIMITATIONS AND RECOMMENDATIONS

Technology plays a significant role in the lives of the 21st-century digital society. The various digital platforms and tools enable ease of access to mediated learning. Nonetheless, incarcerated inmates continue to suffer exclusion from the digital world both for socialization purposes as well as being part of a community of learning. The limited information on how to integrate e-learning for correctional services both in developed and developing countries like Lesotho misinforms the education policy and practice on how to transform the current provision into digital learning.

This paucity is a cause for concern and consideration for future scientific research studies.

E-learning in all forms seeks to transform and revolutionize the traditional face-to-face teaching and learning practice beyond the classroom confines for wider access by marginalized and underserved society groups such as the incarcerated inmates. Through digital platforms and e-learning applications, information sharing, community learning and acquisition of technical skills happen simultaneously in a conducive learning environment independent of space and time restrictions. It promotes independent learning through rigorous self-reflection of the practical application of the theory in real-life practice for complete human development.

In this chapter, the author recommends a curriculum reform that integrates WhatsApp messenger, an instant messaging application as an e-learning tool for facilitating rehabilitation programs offered for incarcerated inmates in Lesotho. It is also suggested that Lesotho taps on the existing collaboration between LCS, CSOs, and NGOs as a strategy to take the integration process further. Some of the identified threats to integration of WhatsApp in learning include learner distraction and lack of a clear cut strategy to inform priority setting between social media communication and study time. These threats need to be explored further in future research. It is also pertinent to explore the resourcefulness of the academia community in Lesotho to support the integration of e-learning in correctional services through scientific discussions and social dialogues. The positive effects of WhatsApp in creating a community of learning and a sense of belong where support, resource sharing learner-directed learning are underlined. Based on the limited research in this area, the findings of this study may also be applicable to catalyze the rehabilitation programs offered for incarcerated inmates in other countries other than Lesotho.

In view of Lesotho's unhealthy economy, the integration of technology in rehabilitation programs may sound like a far-fetched thought. However, the author takes cognizance of the strong social capital networks that LCS has already established with the CSOs and NGOs working in Lesotho. LCS can use this social network as a capital for transforming education provision for inmates from classroom based into a digital provision. At an institutional level, the country can tap on the local academia in the local institutions of higher education to investigate the feasibility of integrating e-learning using mobile device application such as WhatsApp messenger for Lesotho's inmates. The researcher's strong held believe is that WhatsApp messenger will be of great benefit in that the instant messaging facility enables immediate feedback provision benefit and a source of learner motivation. The application will enable the inmates to develop substantial skills and open access to continuing learning beyond their incarceration to enhance their participation in the job market search while also reducing the costs of recidivism for Lesotho.

8.7 SUMMARY

Prison life comes with transformation of the inmates from their comfort zones to adapt into a revolutionized correctional service environment, which has a straight jacket hierarchy and policies governing these institutions. This simultaneously implies a need for behavior and identity change that appreciates deprivation of basic enjoyments such as freedom and limited access to the world outside the institution. The new life also comes with long-term implications on the inmates' resocialization and community re-entry as productive citizens outside prison. Therefore, this chapter advocates for inclusion of WhatsApp messenger as an e-learning tool that connects the inmates with the outside world while also enabling independent learning for smooth transitioning of inmates into reality where access to job market requires ICT skills.

The literature findings revealed that the trending nature of e-learning has taken over from traditional ways of doing and ventured into the digital world, the notion which education provision is not immune inclusive of rehabilitation programs for incarcerated inmates. It was also found that WhatsApp messenger provides a collaborative digital learning where technology becomes the main mode of content delivery and sharing of information. Different scientific fields of study that used WhatsApp messenger for learning reported improved learner performance, confidence, and motivation to learn.

The collaborative nature of WhatsApp messenger through the chat group facility enables affordable access to peer support as well as teacher support while also transforming learning from the individualistic perspective into a communal philosophy where members equally shared the benefits. However, not much is known regarding the feasibility of integrating WhatsApp in Lesotho's rehabilitation programs in terms of what informs the inmates' choices of media platforms, the available technological skills, network coverage, and the cost-effectiveness.

Based on the results in this study, the chapter makes the following recommendations: A need for scientific baseline research studies that will inform the implementation process. It is suggested that Lesotho can take advantage of the existing social capital between LCS, CSOs, and NGOs in Lesotho as leverage for collaborated efforts of technology integration for Basotho inmates. The chapter acknowledges the potential threat of the identified challenges of WhatsApp and recommends a need to explore the limitations further in future research. It is also pertinent to explore the resourcefulness of the academia community in Lesotho to support the integration of e-learning in correctional services through scientific discussions and social dialogues.

KEYWORDS

- **e-learning**
- **inmates**
- **mobile devices**
- **mobile learning**
- **smartphones**
- **social interaction**
- **social media**
- **WhatsApp**

REFERENCES

1. Aburezeq, I. M.; Ishtaiwa, F. F. The Impact of Whatsapp on Interaction in an Arabic Language Teaching Course. *Int. J Arts Sci.* **2013,** *6* (3), 165–180.
2. Adebola, O. A. Semiotics in the WhatsApp Conversations of Undergraduate Students of Obafemi Awolowo University, Ile Ife, Osun State, Nigeria. *Int. J. Engl. Lit.* **2017,** *8* (5), 43–62.
3. Alenazi, A. A. WhatsApp Messenger as a Learning Tool: An Investigation of Pre-service Teachers' Learning Without Instructor Presence. *J. Educ. Train. Stud.* **2018,** *6* (4), 1–8.
4. Awada, G. Effect of WhatsApp on Critique Writing Proficiency and Perceptions Towards Learning. *Cogent Educ.* **2016,** *3,* 1–25.

5. Barrett, M. S.; Barker, J. S. Developing Learning Identities in and Through Music: A Case Study of the Outcomes of a Music Program in an Australian Juvenile Detention Center. *Int. J. Music Educ.* **2012,** *30* (3), 244–259.
6. Behan, C. Learning to escape: Prison Education, Rehabilitation and the Potential for Transformation. *J. Prison Educ Reentry* **2014,** *1* (1), 20–31.
7. Bidin, S.; Ziden, A. A. Adoption and Application of Mobile Learning in the Education Industry. *Procedia Soc. Behav. Sci.* **2013,** *90*, 720–729.
8. Boulos, M. N. K.; Giustini, D. M.; Wheeler, S. Instagram and WhatsApp in Health and Healthcare: An Overview. *Future Internet* **2016,** *8*, 37–49.
9. Brailovskaia, J.; Margraf, J. Comparing Facebook Users and Facebook Non-users: Relationship Between Personality Traits and Mental Health Variables: An exploratory Study. *PLoS One* **2016,** *11* (12), 1–17.
10. Disetha, A.; Eikeland, O.; Terje Mangera, T.; Hetlanda, H. Education Prison for Inmates: Course Experience, Motivation, and Learning Strategies as Indicators of Evaluation. *Educ. Res. Eval.* **2008,** *14* (3), 201–214.
11. Dixon, N. Stranger-ness and Belonging in a Neighborhood Whatsapp Group. *Open Cult. Stud.* **2017,** *1*, 493–503.
12. Dube, S.; Scott, E. A Survey of the University Students' Perspectives About Using Digital Technologies in Education: Zimbabwean Case. *IAFOR J Educ.* **2017,** *5* (1), 125–139.
13. Farley, H.; Pike, A. Engaging Prisoners in Education: Reducing Risk and Recidivism. Advancing Corrections. *J. Int. Correct. Prison. Assoc.* **2016,** *1*, 65–73.
14. Fattah, S. E. A. The Effectiveness of Using Whatsapp Messenger as One of the Mobile Learning Techniques to Develop Students' Writing Skills. *J. Educ. Pract.* **2015,** *6* (32), 115–127.
15. Government of Lesotho. Constitution. Government Printing, Maseru, Lesotho, 1983; p 105. https://www.constituteproject.org/constitution/Lesotho_1998.pdf?lang=en (accessed Aug 15, 2014).
16. Government of Lesotho. Lesotho Education Act No. 3 of 2010: Lesotho Government Gazette ExtraOrdinary LV(20). Government Printing, Maseru, Lesotho, 2011; p 51. https://lesotholii.org/ls/legislation/act/2010/3/education_act_2010.pdf (accessed Sept 16, 2016).
17. Greenhow, C.; Robelia, B. Informal Learning and Identity Formation in Online Social Networks. *Learning, Media Technol.* **2009,** *34* (2), 119–140.
18. Hawley, J.; Murphy, I.; Souto-Otero, M. Prison Education and Training in Europe: Current State-of-play and Challenges. European Commission, Brussels, 2013; p 62. ec.europa.eu/dgs/education_culture/repository/education/library/.../prison_en.pdf (accessed Nov 20, 2017).
19. Hopkins, S.; Farley, H. A Prisoners' Island: Teaching Australian Incarcerated Students in the Digital Age. *J. Prison Educ.* **2014,** *1* (1), 42–51.
20. Jebari, K.; Boussedra, F.; Ettouhami, A. Teaching "Information Systems Management" with Moodle. *iJET* **2017,** *12* (4), 4–16.
21. Jiméneza, E. G.; Garcíab, R. L. Identity of the Inmate and New Spaces of Re-socialization After Leaving the Correctional Institution. *Procedia Soc. Behav. Sci.* **2014,** *132*, 351–356.

22. Johnson, L. R. Cultural and Social Uses of Orality and Functional Literacy: A Narrative Approach. *Read. Writ.* **2016,** *7* (1), 1–7.
23. Kjelsberg, E.; Skoglund, T. H.; Rustad A. B. Attitudes Towards Prisoners, as Reported by Prison Inmates, Prison Employees and College Students. *BMC Public Health* **2007,** *7* (71), 1–9.
24. Montag, C.; Błaszkiewicz, B.; Sariyska, R.; Bernd Lachmann, B.; Andone I,; Boris Trendafilov, B.; Eibes, M.; Markowetz, A. Smartphone Usage in the 21st Century: Who Is Active on WhatsApp? *BMC Res. Notes* **2015,** *8* (331), 1–6.
25. Montoneri, B. Facebook Posts as Complementary Teaching Material for a French University Course in Taiwan. *IAFOR J. Educ.* **2017,** *5* (1), 141–162.
26. Ngozwana, N. Rehabilitating ex-offenders Through Non-formal Education in Lesotho. *IAFOR J. Educ.* **2017,** *5* (1), 111–121.
27. Pérez-Aldeguer, S. Developing a Musical Workshop with Young Inmates at Rehabilitation and Reintegration Enters. *Procedia Soc. Behav. Sci.* **2015,** *191*, 668–672.
28. Raiman, L.; Antbring, R.; Mahmood, A. WhatsApp Messenger as a Tool to Supplement Medical Education for Medical in Clinical Attachment. *BMC Med. Educ.* **2017,** *17* (7), 1–9.
29. Ressing, M.; Blettner, M.; Klug, S. J. Systematic Literature Reviews and Meta-analyses: Part 6 of a Series on Evaluation of Scientific Publications. *Deutsches Ärzteblatt Int.* **2009,** *106* (27), 456–463.
30. Saunders, G.; Oradini, F.; Clements, M. SMART Teaching in New and Old Classrooms. *IAFOR J. Educ.* **2017,** *5* (1), 85–105.
31. Setoi, M. *Youth and Adult Learning and Education in Lesotho.* OSISA, dvv, Bonn, Germany, 2012; p 200. www.osisa.org/...learning/education/lesotho/youth-and-adult-learning-and-education-in_Lesotho (accessed Nov 12, 2014).
32. Shakya, H. B.; Christakis, N. A. Association of Facebook Use with Compromised Well-being: A Longitudinal Study. *Am. J. Epidemiol.* **2017,** *185* (3), 203–211.
33. Snilstveit, B.; Oliver, S.; Vojtkova, M. Narrative Approaches to Systematic Review and Synthesis of Evidence for International Development Policy and Practice. *J. Dev. Eff.* **2012,** *4* (3), 409–429.
34. Stanković, J.; Milovanović, S.; Radović, O. Economic Themes, **2017,** *55* (2), 281–304.
35. Ta'amneh, M. A. A. A. The Effect of Using Whatsapp Messenger in Learning English Language Among University Students. *Int. Res. Educ.* **2017,** *5* (1), 143–151.
36. UNESCO. Policy Guidelines on Inclusion in Education. UNESCO Institute of Lifelong Learning, Paris, France, 2009; p 38. http://unesdoc.unesco.org/images/0017/001778/177849e.pdf (accessed Nov 24, 2017).
37. UNESCO. Accountability in Education: Meeting Our Commitments. Global Monitoring Reports 2017/18, UNESCO, Paris, France, 2017; p 59. http://unesdoc.unesco.org/images/0025/002595/259593e.pdf (accessed Nov 24, 2017).
38. Vaičiūnienė,V.; Romeris, M.; Mažeikienė, V. Social Media in Adult Education: Insights Gained From Grundtvig Learning Partnership Project—Institutional Strategies Targeting the Uptake of Social Networking in Adult Education ISTUS). *Soc. Technol.* **2012,** *2* (2), 473–482.

39. Waheed, M.; Kaur, K.; Ain, N.; Hussain, N. Perceived Learning Outcomes from Moodle: An Empirical Study of Intrinsic and Extrinsic Motivating Factors. *Inf. Dev.* **2016,** *32* (4), 1001–1013.
40. Westrheim, K.; Manger, T. Iraqi Prisoners in Norway: Educational Background, Participation, Preferences and Barriers to Education. *J. Prison Educ. Reentry* **2014,** *1* (1), 6–19.
41. Wilson, R. E.; Gosling, S. D.; Graham, L. T. A Review of Facebook Research in the Social Sciences. *Persp. Psychol. Sci.* **2012,** *7* (3) 203–220.
42. Wright, R. (2014). A Prisoners' Island: Teaching Australian Incarcerated Students in the Digital Age. *J. Prison Educ. Reentry* **2014,** *1* (1), 32–41.
43. Yasunaga, M. (2014). Non-formal Education as a Means to Meet the Learning Needs of Out-of-school Children and Adolescents. UNESCO, Paris, France, 2014; p 295. http://allinschool.org/wp-content/uploads/2015/01/OOSC-2014-Non-formal-educationfor-OOSC-final.pdf (accessed Nov 24, 2017).
44. Yeboah, J.; Ewur, G. D. The Impact of Whatsapp Messenger Usage on Students' Performance in Tertiary Institutions in Ghana. *J. Educ. Pract.* **2014,** *5* (6), 157–164.

CHAPTER 9

Role of Teacher–Developer Relationship in Mobile Apps Learning

SORAYA SEDKAOUI

ABSTRACT

The use of mobile applications is growing exponentially in the area of communications and computing. More powerful, cheaper, and more connected mobile tools have entered the market. In conducting a literature review on the integration of such technology in learning process, the author noticed the existence of several mobile learning applications that may be useful in an educational institution. These applications offer various types of functionality by harnessing the potential of mobile technologies to enrich the learning and teaching experience, as well as achieve targeted learning goals. However, there is a gap between teachers and developers in the development of mobile learning applications. More precisely, the author observed lack of formalism for the functionalities of these applications. Because of the process of creation of such an application, it is necessary to respect not only the technological and computer needs but also the pedagogic requirements and the educational needs. This chapter addresses importance of relationship "teacher–developer" in the creation of mobile applications as a key determinant for enhancing the quality of education and learning. This chapter highlights the importance of such a relationship to adopt technological context with the educational one, and define how building application based on teachers' sources and what they already know about the students' learning needs.

9.1 INTRODUCTION

The advent of information technology (IT) and the Internet has impacted our daily lives. Everyday, people are bounded by several technology devices in every aspect of their daily routines. With rapid diffusion of this kind of technology, many devices are now used in many disciplines. The adult population, in many countries, has more than one mobile device per person.[31]

Mobile technologies keep growing and continue to attract rapidly new users because this technology provides increasing capacity, and enables more sophisticated use. They are used most frequently for web browsing, chatting, getting news, purchasing, and learning. Since they are becoming very accessible for individuals over the world, it can have a significant role in educational development and facilitating learning.

After many years as teacher at the university and as a researcher, the author has noticed the importance of mobile uses which have the potential to contribute in the improvement of education and in exciting the learning movement. From this point of view, mobile technology can be understood as a means for learners to dramatically expand the kind of meaningful information (information model) in context with the total information (information space) available for solving a problem.[20]

In addition, mobile applications for learning cover every mechanism a teacher might want to use for developing a tool for student's learning. However, it should be noticed that the promise of mobile learning is wholly dependent on teacher's integration, participation, and engagement (as educational professionals with excellent educational resources) in the creation and development of applications that will help students to learn and feel a bit more engaged.

Actually, there are plenty of applications for learning in the educational environment today; however, are they designed for fulfilling the needs of learning, especially in today's digitalized world characterized by wide use of smartphones by students? Therefore, there is a reciprocal need between the teacher, as an educational resource, and the developer who has the ability to schematize professional's ideas to support the learning process.

This mutual relationship, between teachers and developers despite the difference in their talents and field, is defined according to our point of view, as a key determinant for making and creating efficient applications learning that accomplish the objectives of educational and learning process.

This chapter highlights an efficient mobile learning that must respond to the significant changes and IT advent challenges in the educational sector in order to operationalize great learning apps. And one way to overcome these challenges is to improve the relationship "teacher–developer," that is, with the contribution of people who understand chat learners need. Developers can get the information they need for designing and creating learning tools and develop quality resources both for teachers and students.

The chapter is organized as follows: The concept of mobile learning is discussed (Section 9.2) by looking at the history of technology and its uses in education. Section 9.3 includes: (1) definition of the concept (Section 9.3) based on the educational and technological perspectives and (2) emphasis on its importance both for the learners and the teachers; and illustration of its various features that can improve the learning process. The development process and implementation of a learning application is discussed in Section 9.4. Section 9.5 highlights its importance by describing the role and the educational needs to develop an effective application.

9.2 LEARNING AND MOBILE TECHNOLOGY

Technology has transformed our world and it has developed so fast that we have difficulty to follow. Nowadays, it is difficult to imagine a world where the technology applications are not omnipresent. Information and communication technology (ICT) applications now exert their influence in almost every field, from banking to politics to education, and they serve to increase productivity in many areas.

First commercialized as communications and entertainment devices, ICTs are now playing an important role in economies and societies at large. After having been widely used in the technical fields, the world of education has taken over this technology in order to adapt it to the learning context. Over the time, the process, that lead to the acquisition of know-how and knowledge, uses and receives more technological tools and components, and this approach has contributed to the development of communication information and the interactive learning context.

Figure 9.1 illustrates that ICT applications continue to expand the boundaries of education into an "anytime/anywhere" experience. This is how these practices in educational context have proved their worth and then have become widespread. It summarizes the brief overview on the chronology of technology integration and their effect on educational environment. This integration has significantly changed and enhanced the learning context since its first use.

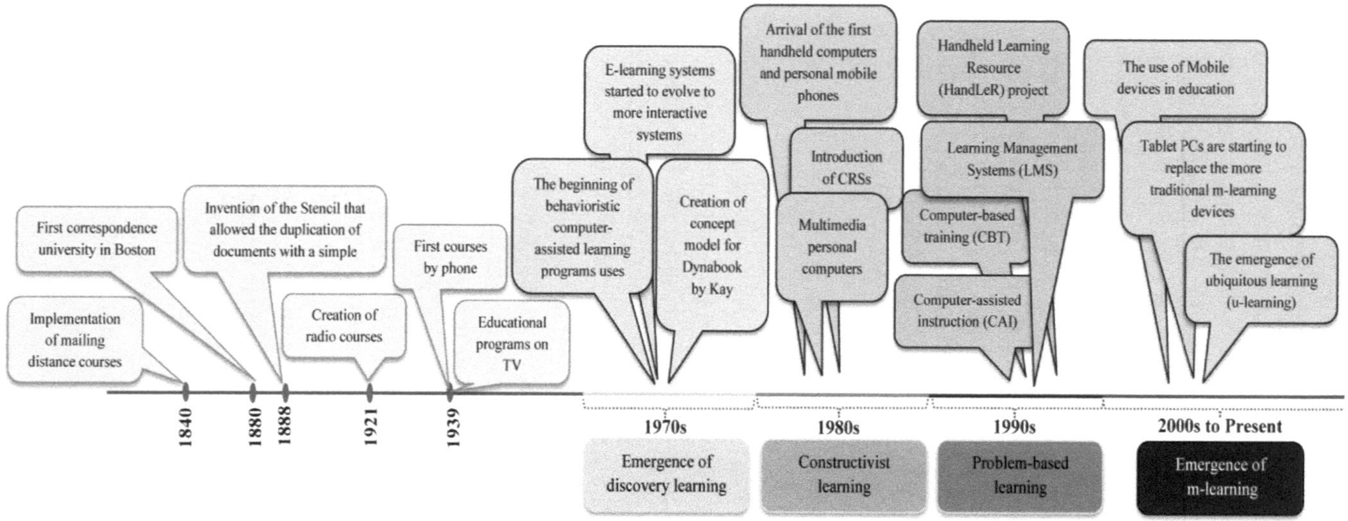

FIGURE 9.1 **(See color insert.)** History of technology integration in learning process.

From the first testing machine, which has been implemented to allow student's testing and practicing of knowledge independently for e-learning and massive open online courses (MOOC) continue to gain popularity in today's learning environment, and from the moving toward mobile learning to the emergence of the ubiquitous learning context. At each stage of the development, educational context has embraced the opportunities offered by technology development to improve learning and adapt it to the current challenges.

It should be noticed that this change is not only about the technologies we use but also the pedagogical methods adopted (teaching), and although obviously, the way students learn. In recent decades, a new way of learning has become increasingly popular due to the evolution and the development of technologies, especially mobile phones. Today mobile learning, as a potential of game-changer,[6] has become the center of interest, attracting the attention of many researchers and scientists around the world, who are interested in its development, its distribution, and how to integrate for better serving the learning process.[7]

Many mobile devices and wireless technologies have been developed. On one hand, their size has decreased and redesigned to better adapt to technological advancements; and the power, the speed, the memory, and functionalities of these technologies have increased. Mobile technologies have made significant progress in its development in the recent years. These have become an integral part of people's daily life.[29]

It is affirmative that we love our smartphones and we use them daily to chat with friends, to get news, and even to learn. Integrating mobile technology to give opportunity to new ways of learning and teaching has opened the field to researchers. Mobile learning, in this context, can be considered as an act determined by the learner because he uses and appeals these technologies to enhance and develop his knowledge and capabilities in order to support the aims behind the learning process.[20]

Many research studies, which aimed to analyze the efficiency of the mobile technology in learning environment, have shown that the use of such technologies participate certainly in the learners learning process, by helping to be more engaged and promoting the learning results.[1,12,28] The evolution of mobile technologies as well as the diversity of mobile devices and their characteristics have favored their exploitation in the field of education.

The use of mobile devices is a growing interest among many researchers and some researchers treat it as key element in the implementation of digital learning and for the advanced model of e-learning.[30] Indeed, there are several mobile learning applications offering multitude of features.

9.3 MOBILE LEARNING: APPLICATIONS AND PEDAGOGY TRANSFORMATION

Before highlighting the feature offered by mobile applications in an educational context and addressing the pedagogical shift and the teaching approaches, the developer should understand what mobile learning is. For this purpose, we will lean on two aspects, for example, technological and educational aspects.

From the technological point of view, mobile learning can be considered as learning that takes place using mobile devices.[25] The mobile learning in an education process is any learning activity that can be achieved through mobile devices, or in premises where mobile equipment is available.[8] Also mobile learning or any educational provision is dominant technologies using portable devices.[34]

Mobile learning from educational perspective focuses on the learner, rather than physical mobile devices. Mobile learning is therefore the experience and opportunity offered by the evolution of pedagogical technologies to foster the creation of knowledge, satisfaction of curiosity, and collaboration between learners. Such learning is acquired anytime and anywhere at the request of the learner.[21] The mobile learning is also any type of learning that takes place when the learner is not in a fixed or predetermined place, or it is the learning that takes place when the learner benefits from the field of opportunity that is offered by these mobile technologies.[23]

Therefore, the technological aspect is reflected in the fact that the features to be specified and formalized are offered on mobile devices. As for the pedagogical aspect, it is the learner (as a user of the proposed mobile application), who must be able to fully benefit from the various features offered.

The technological aspect, therefore, defines the collaboration between technology and mobile networks, to facilitate, support, improve, and extend the reach of teaching and learning. While the educational aspect characterizes mobile learning as being highly situated, personal, collaborative, and in the long-term, learning is really learner centered.[24]

Indeed, the effectiveness of what a learner (or student) can learn and how he learns is related to his motivation.[16] And given their characteristics (connectivity, familiarity, personalization, mobile devices are invaluable source of motivation.

Although mobile learning can stimulate motivation of the learner, yet mobile devices have not been specifically designed for education[33] according to some authors. Also some physical characters on mobile devices, such as the small screen size with limited input capabilities, may reduce the usability of mobile devices.[17]

However, this does not hide the importance of mobile applications and its various features that have improved the learning quality by supporting the teachers in their educational tasks and by promoting a more suitable learning environment to the student's needs. The mobile applications' showing supporting pedagogy that we have been able to draw from our literature review can be categorized as follows:

- For learning, following six features were identified:
  - Note taking: Many students reported using the iPad as a mobile device for taking text or voice notes during class sessions.[2,5] For this purpose, they use some applications such as "Evernote," "Poplet," and "inClass" They can also dictate their notes to the device, which transcribes them in software via for example the "Dragon Dictation application."
- Documents presentation: The learner can use mobile devices to create slides. The use of mobile device to create presentations makes the course session more motivating.
- Reading documents: Access to course notes or presentations is often done online, either before or after the course session.
- Evaluation: Student can participate in advancing the learning process by evaluating his own work. An experiment was conducted with PDAs, developing a mobile appraisal participation system. Students were able to use this system to evaluate presentations and to self-evaluate as well.
- Interaction and collaboration: The interaction is considered to be a complete loop of messages from and to the learner. The reciprocal participant may be a teacher or another learner. The loop occurs regardless of the technology or means of communication.[19] When it comes to collaboration, a collaborative learning environment can

be defined by an environment that combines the social elements and building blocks of the learning process while using integrated technologies to support both.[15]

- Exploration: One of the important features of mobile devices is connectivity. It allows students to access online resources by browsing the Internet. An example of the importance of using mobile devices for exploration and research is Indiana University, when a journalism student claimed that the use of the University Library's mobile website has facilitated the discovery of new and diverse information.[27]

- For teaching:
  - Evaluation: The teacher can evaluate his students through the use of quizzes. These are considered an ideal way to illustrate and model the most difficult concepts present in a course. An example of a mobile application allowing the teacher to evaluate his students is "Socrative."[35] This application allows teacher to create multiple choice questions, true or false questions, and also short answer questions.
  - Diffusion and content management: The courses to be taught can come in different forms. For example, we notice the frequent use of video podcasting. It is about communicating via visual material, which helps to simulate the learner's senses. In addition, it is a way to contextualize abstract concepts.[11] Among the podcasting delivery experiences, there is a first example of "Duke University" in the US. This university offered in August 2004 to more than 1600 new students of "Ipod." The mobile device was used for various purposes such as broadcasting course content and recording lessons and class discussions.[4]
  - Collaboration and interaction: Some mobile applications provide collaboration and interaction between teachers themselves or between teachers and students. Among these applications is the "Edmodo" application, which allows the teacher to collaborate with his students on projects and share information through a private communication network.[26]
  - Correction of work and assignment of notes: Students can send their work electronically in pdf format on IPad. For example, the teacher can use a mobile application to correct these assignments without having to print them. He can annotate them,

insert images, he can even record audio sequences. Finally, the assigned note can be returned to the student by an email directly from the used application.[18]

9.4 DEVELOPMENT OF MOBILE LEARNING APPLICATION

In all spheres of activity, one always tries to follow the technological evolution, and the world of education cannot escape it. The development of mobile applications is exploding to attend educational field due to the widespread use of mobile devices. Today's learners are characterized by their level of knowledge of technology as a tool. From the young age, they know that the Internet is a fascinating world, they know that touching a screen allows them to penetrate a universe of discoveries, that their creative efforts are just a click away, and that by exploring different types content, they can determine their passions and preferences.

We consider the "Five Moments of Need," which was designed that can be used to make the state of the different educational needs. According to this tool, learners may systematically need help in the following situations:

i) When their environment changes (change).
ii) When they become familiar with the content for the first time (new).
iii) When they have to solve a problem, or fix something that went wrong (solve).
iv) When they want to deepen a subject (plus).
v) When they want to practice, or remember, something (apply).

Above model is organized in two stages; the first two steps are related to obtaining knowledge during a traditional type of training (in class or online). In contrast, the next three steps are related more to performance support. They are designed to help learners reuse their knowledge in the context of their workflow. Thus, an effective learning and performance applications creation strategy must take into account both measures adapted to the learner's integration of content (as part of a classical training) and, more importantly, reuse this knowledge.

Therefore, the first step is to understand how to go about it if one wants to build a mobile learning application. We can of course develop our own application after defining the specifications. In this case, one must go through all stages of the development process. Second way is to rely on the existing or published mobile learning applications, including the architectures that are proposed. An important part of the developer activity also concerns the design, the regular update of the application, its reconfiguration, its maintainability, and its evolution.

Also, in the creation of a mobile learning application, we must consider that the student is not going to sit in front of his phone and scroll through 50 presentation slides or watch a 1 h video. This means that the application content has to be concise and precise.

However, it is not only about the development in the mobile learning applications creation process. We need to recognize that good graphics, design, and programming are not enough. It is important to integrate educational professionals as a stakeholder, in the sense that it holds the knowledge that the developer needs and understand the student's learning needs.

There are following four considerations in the development of a mobile application:

i) Development of a master plan that sweeps the fields of multimedia design and pedagogical engineering through the guiding prism and validating teaching objectives.
ii) Insert this scheme into a more complete integration approach[3] following the precepts of an "Integrative Learning Design Framework (ILDF)." An approach centered on the development of the project architecture must be practical, both oriented toward the learning process and the theory, while being of an empirical nature through repeated interventions taking into account the returns of the learners.
iii) Include the learner in the modeling of the project, especially in the dimensions of exploration and exploitation of contents. The learner must be able to have additional content, not requiring too much cognitive load or a heavy personal investment. Thus, formatted or didacticized resources (except those of the web that are otherwise) are the point of entry of the learner–user. The federating element is here formalized by the proposed macro-tasks that will structure, through an organization of knowledge prescribed by the theme addressed in the program, the emergence of needs, and the exploitation work to initiate according to a fragmented temporality.
iv) Consider how this additional microdevice can fit into a pre-existing macrodevice (and that has more different ways and developed). The mobile application will thus be superimposed as a complementary tool at the level of the mediated content axis and offer

its own interaction membranes on a theme similar to the content proposed in the main interface. Developments to be considered in terms of technology, improvement of programming languages, and the equipment rate of learners' point to a more invasive insertion of mobile applications into existing systems.

The development of mobile learning application is a relatively complex project that requires lot of organization and complementary technical skills. Indeed, it is necessary to respect the graphic charter, the marketing, the communication, etc. Also the fact—that engaging teachers and educational professionals from the concept stage in the development of new applications—will be the way they create a point of difference. It is a team effort, which brings together according to its complexity a number of developers with specific skills and which draws on the skills and knowledge of those who know learning and education best.

The importance effectiveness of any mobile applications must be measured regarding its importance (the why?), every major innovation, the biggest confusion lies in the exact scope (what? and who?), and its implementation (how? and for whom?). If one cannot find the way for its conception "how" and commercialization "for whom," then it will be difficult and ambiguous to understand "what" is this technology, "who" must concretize it and "why" it is so important.

Learners have the possibility to learn everywhere, at any time, whatever they want, and can learn using mobile anywhere: waiting on the train station, the bus, at recess, when they have finished their work, and so on. Many mobile applications for learning are being developed (mathematics, languages, etc.). What is more tempting: instead of playing on phone, learners will be able to use their time to do something productive! Therefore, applications for developers must consider all these aspects to turn the corner on the attractiveness of application.

9.5 TEACHER–DEVELOPER: THE IMPORTANCE OF COLLABORATION

The widespread use of mobile devices (such as smartphones, IPods, and tablets) has accelerated the development of mobile applications. If developers are going to create any learning apps, then understanding the kinds of learning is needed to help learners or students develop their knowledge in such a way that will serve all teachers. They must start by formulating a theory of how professional learning works to influence student learning process, taking into account the several aspects discussed before.

According to the Oxford dictionary, developer is a: "person or thing that develops something." A developer is also known as a "software developer," "computer programmer," "programmer," "software coder," or "software engineer." A mobile application developer develops application for mobile devices (smartphones, tablets). It may also have the mission of optimizing already created applications or adapting websites to mobile support. This professional person can specialize to work on applications dedicated to Android or iOS or even offer applications compatible with many of these operating systems.

The developer must have a solid knowledge of the mobile world, current standards, programming languages (IT context). Algorithm, mathematics, and geometry are also areas that the application developer must master. It must also demonstrate its ability to adapt to understand the needs of the student or the learner and translate them into automatic way.

The developer is the artist and the application creator. The mission is to create technically a mobile application and ensure its computer development. A developer can intervene from A to Z to design the application or to improve those which already exist.

For a developer, who will design and develop a mobile application learning in order to promote it to students, teachers and educational institutions as an educational tools supporting the learning process, he needs the help of educational professionals, who really understand the students' needs and the importance of pedagogy (Fig. 9.2).

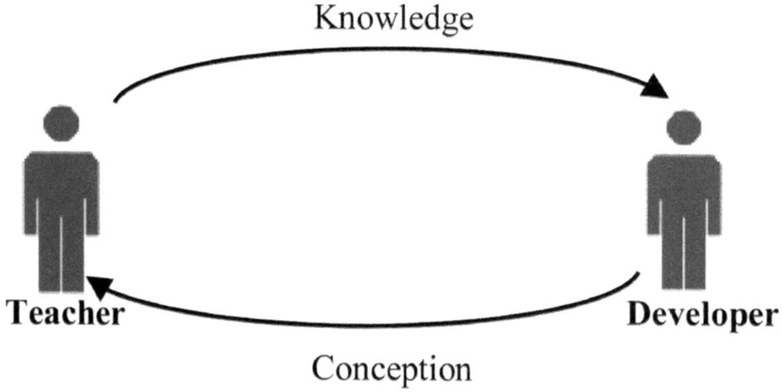

FIGURE 9.2 (See color insert.) The teacher–developer relationship.

It is important to note that one cannot produce learning and educational application designed for mobile tools independently from the systems they intend to be used (educative context) in and without the input of professionals, who understand learning needs and have experience in planning, designing, and developing mobile learning apps. Teacher is a principal actor in the educational context that constituted the learning environment of each student or learner, and integrating this actor in the application creation process may influence how new methods would be translated into mobile practice because of:

- Teachers are present and available during the course;
- They know the facility and what the infrastructure needs to provide;
- They can intervene immediately if there are any difficulties.

This can help the developer to think its application differently not only according to the means of interaction, the size of the screen, the bandwidth but also according to the real needs on the ground. Taken together, these two sources provide a powerful framework for thinking about how best to develop learning applications. The relationship teacher–developer do not only refer to the collaboration between the two but also between the two context (IT context and educative context) in which each one operates (Fig. 9.3).

However, it should be noticed that process of creation of any application cannot be valuable and usable without including the learners in the heart of this process. Because the learner or the student is positioned, as illustrated in the (Fig. 9.4) at the crossroads of the two contexts.

The teacher must play the role of the learner's full program monitor (agenda) who is supposed to use the application; the working time using this application; and informing him about the new advancement and available version of the application, also the teacher can encourage the student to use this application, and so on.

For the preparation of the design of the application, its planning and the educational content are done by the developer with the assistance of teachers, in order to develop the content and facilitate the features and uses. The main objectives through this collaboration are:

i) *Help* the students or learners to overcome the difficulties in choosing the suitable learning methods.

ii) *Conceptualize* and prepare the applications based on the learner's level and needs.
iii) *Identify* the effectiveness of teaching pedagogy, etc.

This implies that organization of knowledge and fundamentals facilitates retrieval and application.

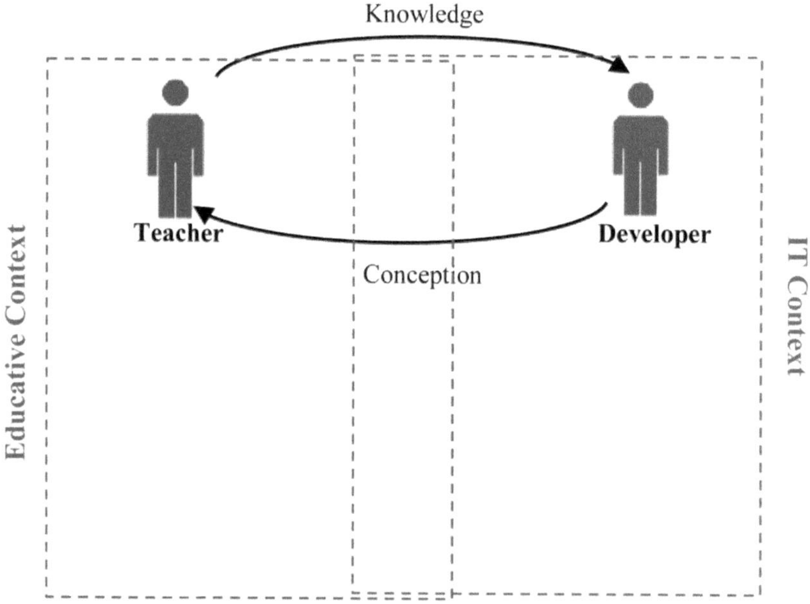

FIGURE 9.3 **(See color insert.)** The extension of the relationship.

After having created the computer programs of the application and having defined the algorithms allowing the treatment of the data, numerous tests of the application are required to ensure at the same time the respect of the educational contents and the good functioning of his creation. In the development application process, developers and teachers must take into account that students or learners need more than the existing methods. It implies that they need more up-to-date and easy technological tools to help them and facilitate the access to available information. This is utmost necessary for them, because it improves their knowledge, especially if they can do so at any time and help them to better manage their time. This is indispensable for students.[9]

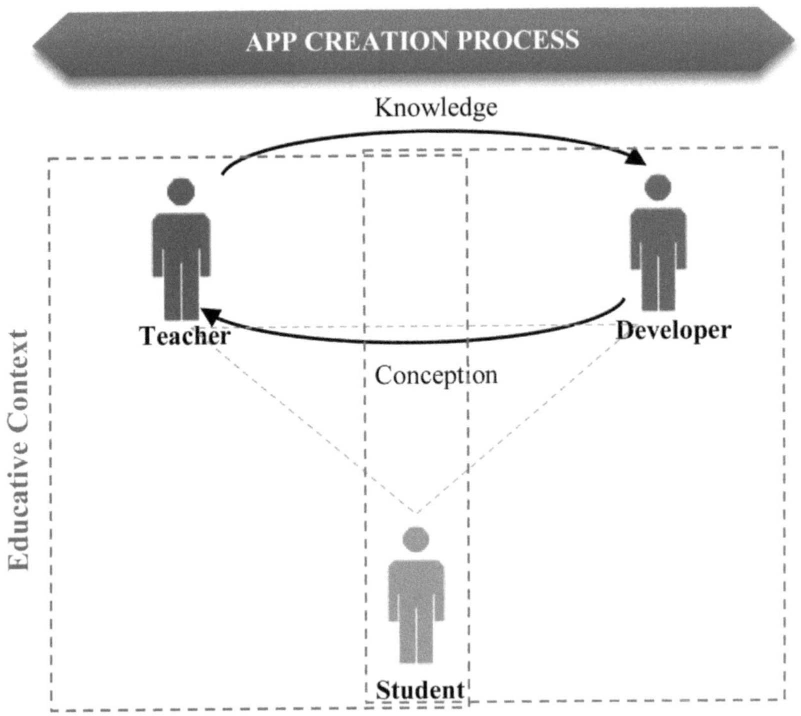

FIGURE 9.4 (See color insert.) The student at the heart of the process.

Another point that should be considered is that student is the one, who move (he is mobile rather than the technology), then the application should therefore not only focus on the design but also on the interaction between learning and technology, while respecting the privacy of each student. Also, considering the relationship between the student and their teacher, this is very important to guide learners understand the advanced technology, help them to be more familiarized with learning program, and be able to apply it to practical educational scenarios in basic education.

Students, for example, can inform their teachers about the difficulties encountered when using the application. In turn, the teachers will transmit the problems to the developer so that he can then improve the application and facilitate its use. This learning context aims to provide a more comprehensive learning process than the traditional approaches and environments requiring communication and technology, in order to facilitate the process and to help this new generation of learners, who are invaded by these technologies to learn[20]

Therefore, this relationship between the learner and developer is very important for validating the architectural model of the application because the teacher and the developer are responsible for the efficiency and quality of the application in question, its content, and features. As the relationship between technical and pedagogical becomes closer, mobile technology will play a role that is both well-defined and increasingly essential within the global educational ecosystem.

9.6 RECOMMENDATIONS

The learning process changes over time and welcomes more factors, which contribute to its development, such as information technology, mobile devices, and other potentials that make the learning process more exciting and amazing. The rapid development of web technology gives more opportunities to develop interactive applications to help students and promote their learning process.

The concept of mobile learning is provoking a real and growing interest. It has also become a buzzword, as evidenced by the illustration in Appendix A, derived from Google Trends. This trend illustrates that the query volume on mobile learning or the learning app is growing rapidly. Mobile technology offers a new space of life, exchange, and interaction. The mobile brings interactivity to the training content necessary, make it alive and federate a community of learners.

In this chapter, the author explored the field of mobile learning, which can be considered as an evolving field. Mobile learning is a real revolution, comparable to the invention of electricity. Some studies have shown the effects of mobile devices and mobile learning on learning and the different skills of students.[10,13,14,32] Using the mobile devices with easy access to internet, students can interact with their teachers from anywhere and can also exchange information with other students.

Throughout this chapter, the author focused to highlight the importance of both the developer and teacher to ensure the effectiveness of any application intended for learning and improving the features offered. To develop any learning mobile application, the developers are brought to work in collaboration with teachers on concrete projects, fun and intuitive to take advantage of current and future software trends: mobility, social networks, usability and accessibility, and the educational context: pedagogy, methodology, needs, etc.

By taking all these factors and criteria into consideration, developer put these into practice by creating a suitable application, which provides the ability for students and teachers to communicate and exchange educational material through the use of the application from their mobile devices. Author affirms that the teacher–developer relationship improves the development process of the application learning and makes it fully operational within the education system because both have much to offer by greatly adding to and evolving the practice of the applications.

9.7 SUMMARY

The mobile device creates new uses and upsets our relationship with the world. It has not only become a new privileged space available anywhere, anytime to interact with others, to play, to learn or to work but also creates new ways to interact with our environment and crystallizes unprecedented opportunities. What roles will mobile play in the transmission of knowledge, a central issue in the development of our civilizations? The rapid emergence of mobile technologies, as well as their considerable contributions in teaching and learning, has fostered their uses to promote education and enrich the educational experience for both learners and teachers. In this chapter, author explored the field of mobile learning; our interest on the importance of integrating the teacher into the process of developing the functionalities of any applications learning. The overall goal of this chapter is to emphasize the importance of developing mobile applications for learning, and to implement the need for an engagement of the teacher (educational world) and the developer (IT). This can enrich the formats and the content of the application learning and contribute in its adaption to the educational context. Marshall-McLuhan in his book, "Understanding Media: The Extensions of the Man (ISBN-13-978-0262631594) indicated that the technologies and tools themselves shape the forms and magnitude of human interactions and associations. The teacher–developer relationship opens the door to other worlds and helps to better study, analyze, and structure the way learners use mobile technologies in different learning contexts.

KEYWORDS

- **mobile applications**
- **mobile learning**
- **mobile technology**
- **student learning**

REFERENCES

1. Ako-Nai, F.; Tan, Q. Location-Based Learning Management System for Adaptive Mobile Learning. *Int. J Information Education Technol.* **2013,** *3* (5), 529–535.
2. Alyahya, S.; Gall, J. E. IPads in Education: A Qualitative Study of Students' Attitudes and Experiences. In *Proceedings of EdMedia: World Conference on Educational Multimedia, Hypermedia and Telecommunications*; Amiel, T., Wilson, B., Eds.; Association for the Advancement of Computing in Education (AACE), Denver, Colorado, USA, 2012, pp 1266–1271.
3. Bannan, B. The Integrative Learning Design Framework: An Illustrated Example from The Domain of Instructional Technology. In: *An Introduction to Educational Design Research*; Plomp, T., Nieveen, N., Eds; SLO: Netherlands Institute for Curriculum Development, Netherlands, 2009, pp 53–73.
4. Belanger, Y. ipod First Year Experience Final Evaluation Report. Technical Report; Duke University, Durham, North Carolina; 2005; p 43. www.educause.edu/ir/library/powerpoint/EDU05149.pps (accessed Jun 30, 2018).
5. Cochrane, T.; Narayan, V.; Oldfield, J. ipadagogy: Appropriating the iPad Within Pedagogical Contexts. *Int. J. Mobile Learning Org.* **2013,** *7* (1), 48–65.
6. Crompton, H.; Traxler, J. *Mobile Learning and Higher Education: Challenges in Context*; Taylor & Francis Group: New York, 2018; p 222.
7. Cumming, T. M. *Sustaining Mobile Learning: Theory, Research, and Practice,* 1st ed.; Routledge: New York, 2016; p 238.
8. Colazzo, L.; Molinari, A.; Ronchetti, M.; Trifonova, A. Towards a Multi-vendor Mobile Learning Management System. In World Conference on E-Learning, 2013. https://www.researchgate.net/publication/228583404 (accessed June 30, 2018).
9. Ekmekçi, E. Integrating Edmodo into Foreign Language Classes as an Assessment Tool. *Participatory Educ. Res. (PER)* **2016,** *1*, 1–11.
10. Fabian, K.; Topping, K. J.; Barron, I. G. Mobile Technology and Mathematics: Effects on Students' Attitudes, Engagement, and Achievement. *J. Comput. Educ.* **2016,** *3* (1), 77–104.
11. Hill, J. L.; Nelson, A. New Technology, New Pedagogy? Employing Video Podcasts in Learning and Teaching About Exotic Ecosystems. *Environ. Educ. Res.* **2011,** *17* (3), 393–408.

12. Huang, R.; Kinshuk, J.; Spector, J. M. *Reshaping Learning: Frontiers of Learning Technology in a Global Context;* Springer Verlag: New York, 2013; p 450.
13. Inel Ekici, D. The Use of Edmodo in Creating An Online Learning Community of Practice for Learning to Teach Science. *Malaysian Online J. Educ. Sci.* **2017,** *5* (2), 91–106.
14. Kuznekoff, J. H.; Titsworth, S. The Impact of Mobile Phone Usage on Student Learning. *Comm. Educ.* **2013,** *62* (3), 233–252.
15. Laurillard, D. The Pedagogical Challenges to Collaborative Technologies. *Int. J. Comput.-Support. Collab. Learn.* **2009,** *4* (1), 5–20.
16. Malone T.W.; Lepper, M.R. Making Learning Fun: A Taxonomy of Intrinsic Motivations for Learning. *Aptit. Learn. Instr.* **1987,** *3* (1987), 223–253.
17. Maniar, N. M-learning to Teach University Students. In *EdMedia*: *World Conference on Educational Media and Technology*; Association for the Advancement of Computing in Education (AACE), Chesapeake, 2007, pp 881–887.
18. Manuguerra, M.; Petocz, P. Promoting Student Engagement by Integrating New Technology into Tertiary Education: The Role of the iPad. *Asian Soc. Sci.* **2011,** *7* (11), 61–85.
19. Markett, C.; Sánchez, A.I.; Weber, S.; Tangney, B. Using Short Message Service to Encourage Interactivity in the Classroom. *Comput. Educ.* **2006,** *46* (3), 280–293.
20. Miller, C.; Doering, A. *The New Landscape of Mobile Learning: Redeigning Education in an App-based World;* Taylor & Francis Group: New York, 2014; p 358.
21. McQuiggan, S.; McQuiggan, J.; Sabourin, J.; Kosturko, L. *Mobile Learning: A Handbook for Developers, Educators, and Learners*; John Wiley & Sons: New York, 2015; p 400.
22. Mokhtar, F. A. Rethinking Conventional Teaching in Language Learning and Proposing Edmodo as Intervention: A Qualitative Analysis. *Malaysian Online J. Educ. Technol.* **2016,** *4* (2), 22–37.
23. O'Malley, C.; Vavoula, G.; Glew, J.; Taylor, J.; Sharples, M.; Lefrere, P.; Lonsdale, P.; Naismith, L.; Waycott, J. Guidelines for Learning/Teaching/Tutoring in a Mobile Environment. Public deliverable from the MOBILearn project (D.4.1), 2005, p 84. https://hal.archives-ouvertes.fr/hal-00696244 (accessed Jun 30, 2018).
24. Paine Schofield, C.; West, T.; Taylor, E. Going Mobile in Executive Education. Technical report; Ashridge Business School, Hertfordshire, United Kingdom, 2011, p 64. http://www.uniconexed.org/wp-content/uploads/2016/10/UNICON-Going_Mobile_In_Executive_Education-Schofield-Taylor-West-Nov-2011.pdf (accessed Jun 30, 2018).
25. Quinn, C. M-learning: Mobile, Wireless, in Your Pocket Learning, 2000. http://www.linezine.com/2.1/features/cqmmwiyp.htm (accessed Jan 16, 2018).
26. Robledo, S. J. Mobile Devices for Learning: What You Need to Know. George Lucas Educational Foundation : San Rafael, CA 2012. https://eric.ed.gov/?id=ED539398 (accessed Jun 30, 2018).
27. Rossing, J. P.; Miller, W. M.; Cecil, A. K.; Stamper, S. E. ilearning: The Future Of Higher Education? Student Perceptions on Learning with Mobile Tablets. *J. Scholarsh. Teach. Learn.* **2012,** *12* (2), 1–26.
28. Sampson, D.G.; Isaias, P.; Ifenthaler, D.; Spector, J.M. Context-aware Adaptive and Personalized Mobile Learning Systems. In: *Ubiquitous and Mobile Learning in the Digital Age*; Sampson, D. G., Isaias, P., Ifenthaler, D., Spector, J. M., Eds.; Springer Publishing Company: New York, 2013; pp 3–17.

29. Sarwar, M.; Soomro, T.R. Impact of Smartphone's on Society. *Eur J. Sci. Res.* **2013,** *98* (2), 216–226.
30. Shih, J.L.; Chu, H.C.; Hwang, G.J. An Investigation of Attitudes of Students and Teachers About Participating in a Context-aware Ubiquitous Learning Activity. *British J. Educ. Technol.* **2011,** *42* (3), 373–394.
31. Statista. Share of Mobile Device Owners Worldwide from 2011 to 2016, by Number of Device Owned, 2016. https://www.statista.com/statistics/245501/multiple-device-ownership-worldwide/ (accessed Feb 8, 2018).
32. Sung, Y. T.; Chang, K. E.; Liu, T. C. The Effects of Integrating Mobile Devices With Teaching and Learning on Students' Learning Performance: A Meta-analysis and Research Synthesis. *Comput. Educ.* **2016,** *94*, 252–275.
33. Traxler, J. Students and Mobile Devices. *Res. Learn. Technol.* **2010,** *18* (2), 149–160.
34. Traxler, J. Defining Mobile Learning. In *IADIS International Conference Mobile Learning*; IADIS, 2005; pp 261–266. http://www.academia.edu/2810810/Defining_mobile_learning (accessed Feb 8, 2018).
35. Wash, P. D. Taking Advantage of Mobile Devices: Using Socrative in the Classroom. *J. Teach. Learn. Technol.* **2014,** *3* (1), 99–101.

APPENDIX A

Google research volume around mobile learning and learning app.

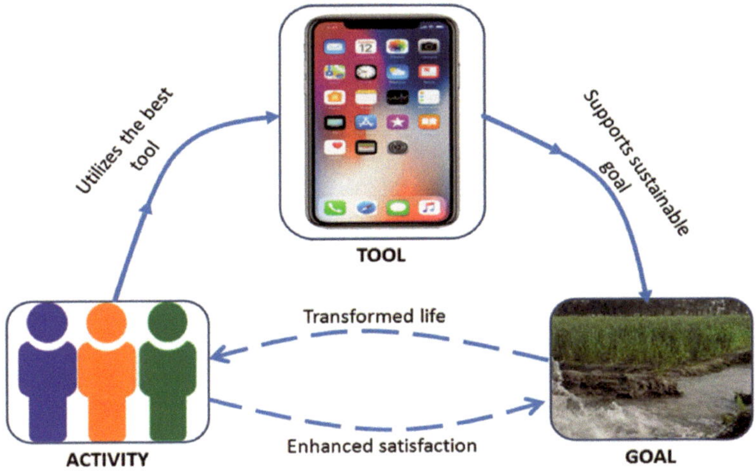

FIGURE 1.1 The conceived nexus of FFO activities, the tool and goal.

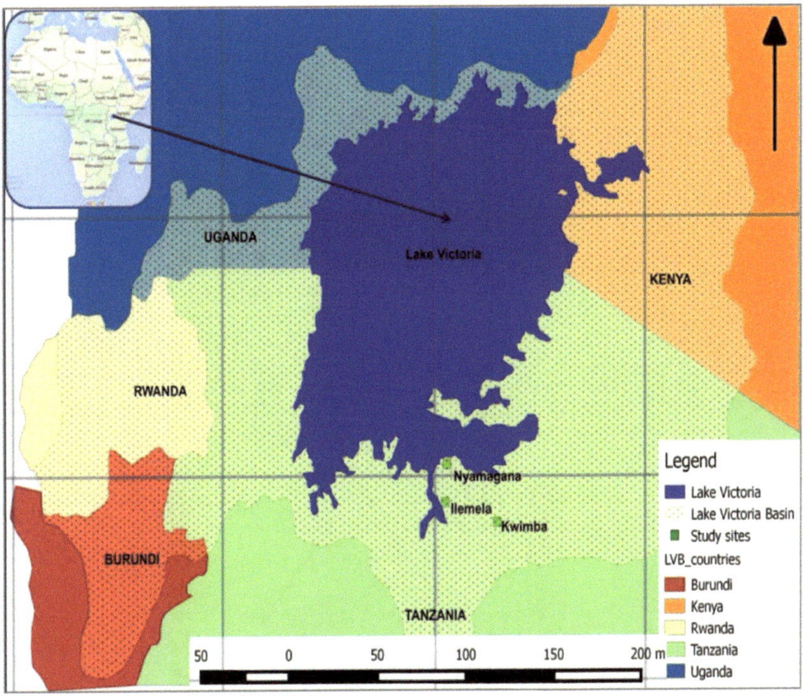

FIGURE 1.2 Map of LVB showing the location of study sites in Tanzania.

Source: Authors' creation with Google Maps.

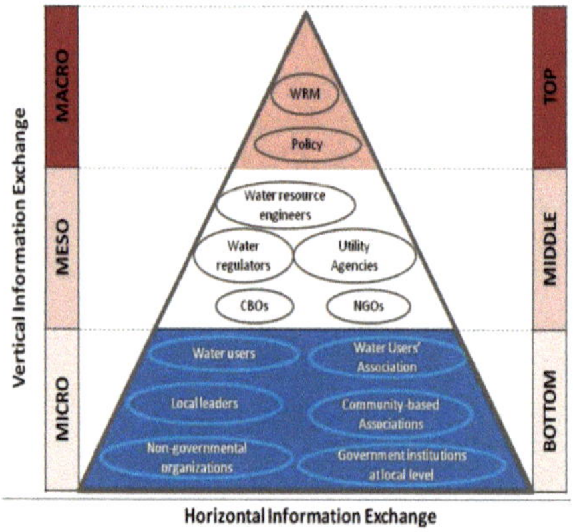

FIGURE 1.3 The VIE and HIE for the three scales of WRM.

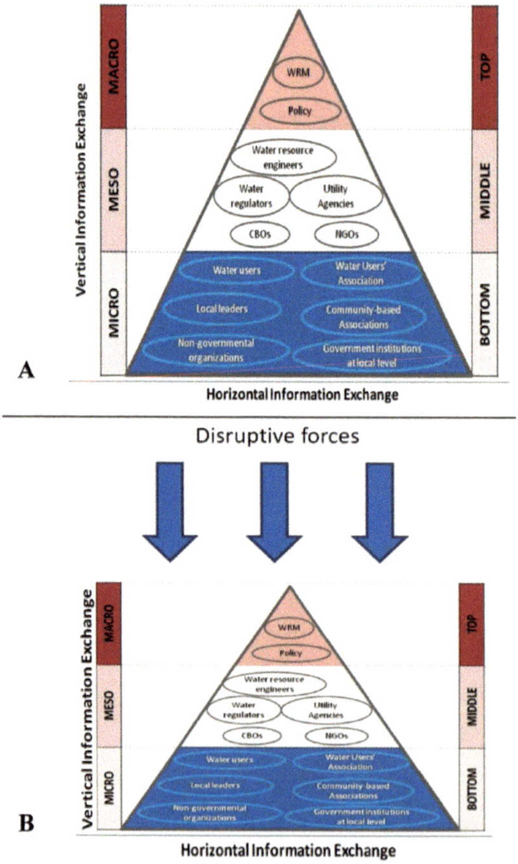

FIGURE 1.4 The potential action of disruptive forces on the vertical information exchange: before (A), after (B).

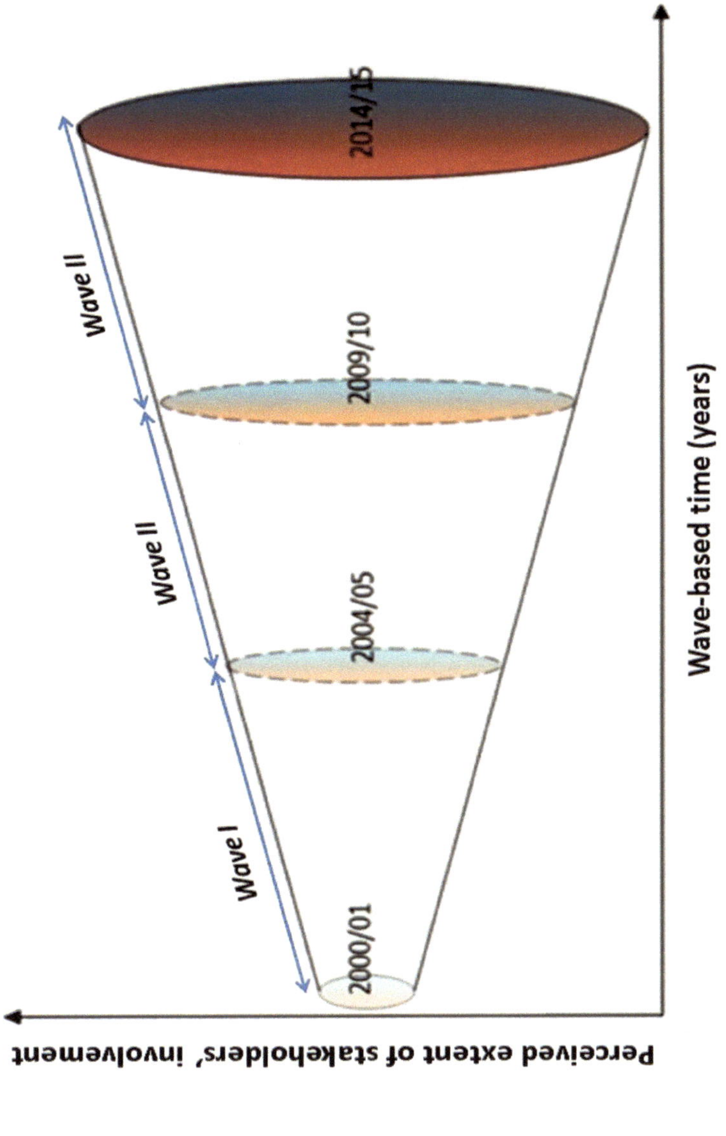

FIGURE 1.5 Trends and aggregation of disruptive forces during different waves.

Source: Adapted with permission from Ref. [32]

FIGURE 5.1 Before preprocessing showing infected leaf.

FIGURE 5.2 After preprocessing.

FIGURE 5.3 Before preprocessing showing infected stem.

FIGURE 5.4 After preprocessing.

FIGURE 5.5 Before preprocessing showing infected fruit.

FIGURE 5.6 After preprocessing.

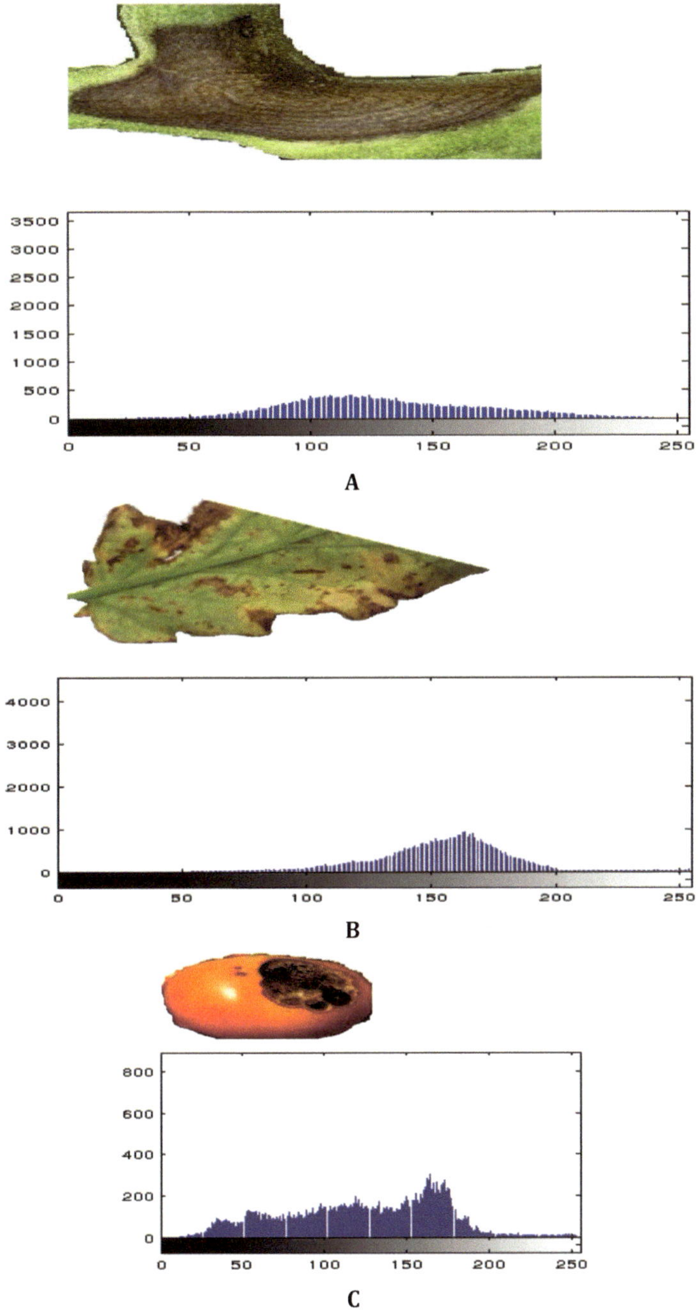

FIGURE 5.8 Feature extraction.

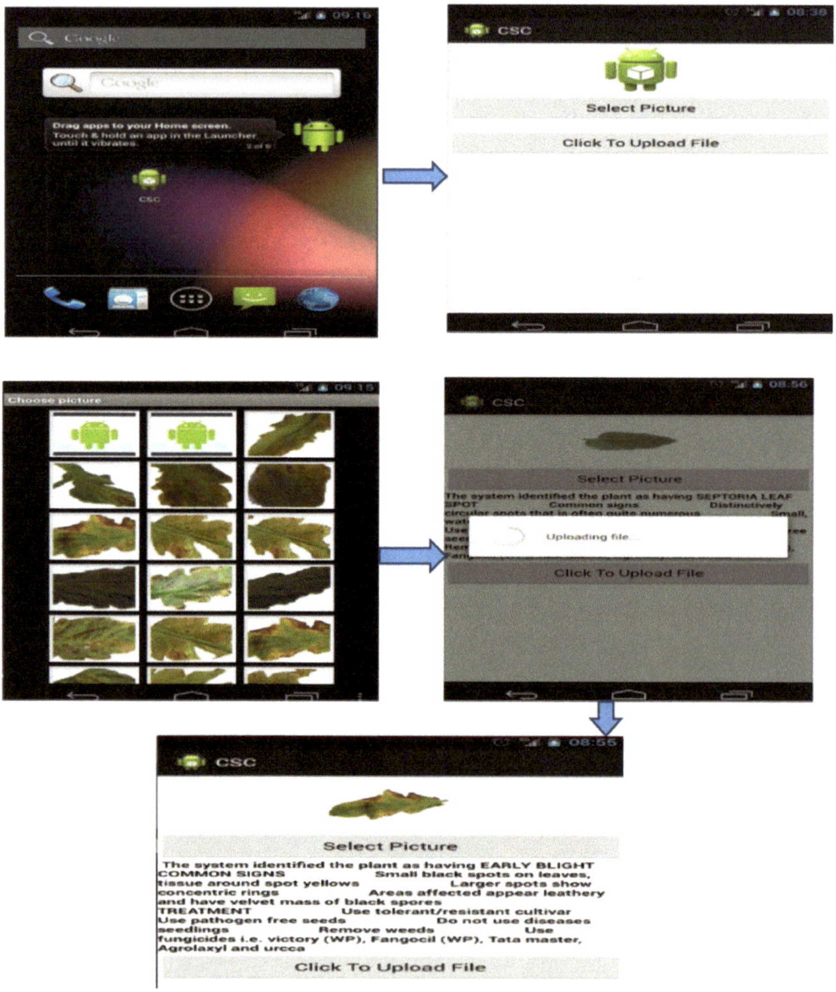

FIGURE 5.10 The diagnosis process.

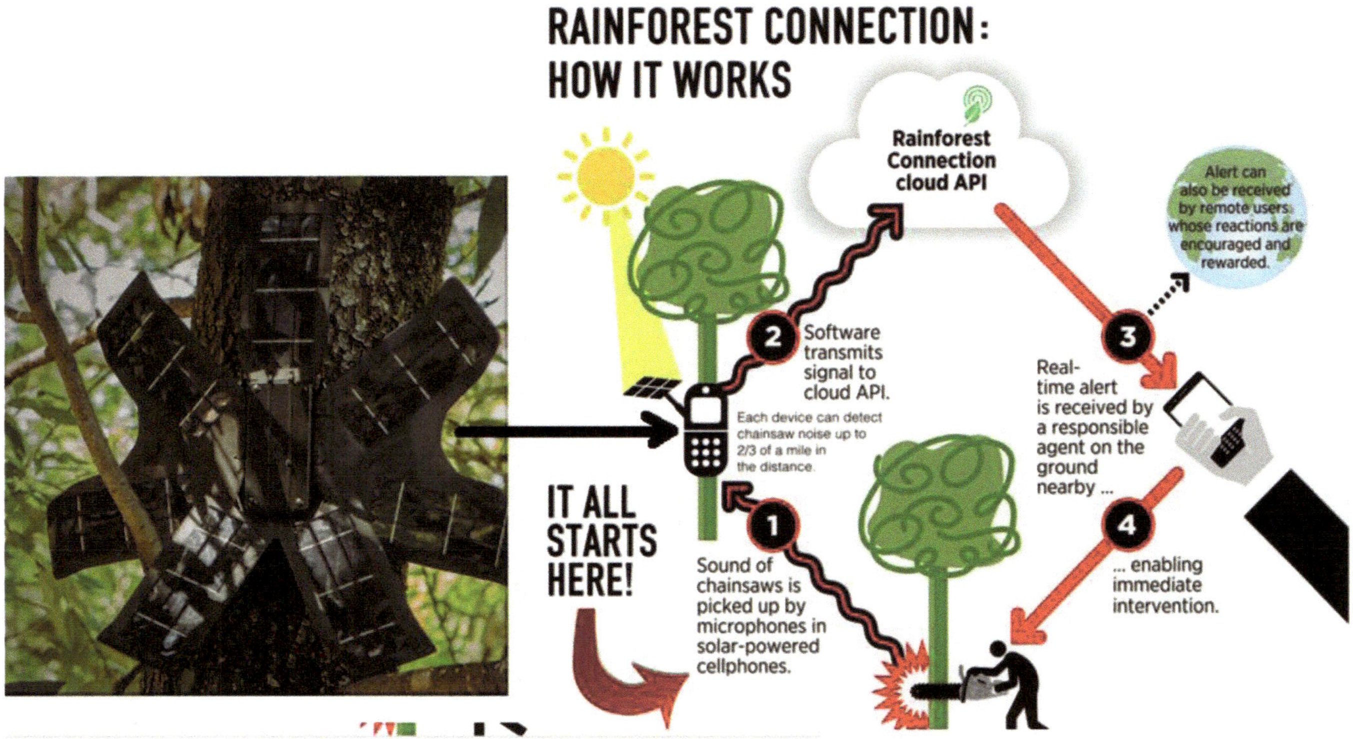

FIGURE 6.1 RFCx device and its network connection

Source: Adapted with permission from Ref. [7].

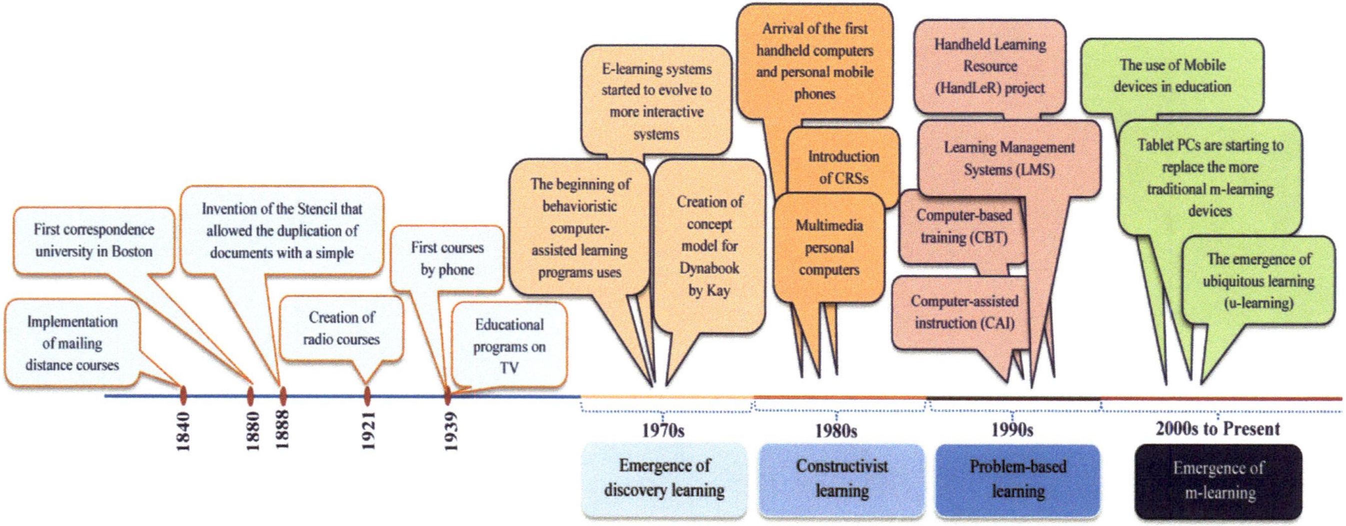

FIGURE 9.1 History of technology integration in learning process.

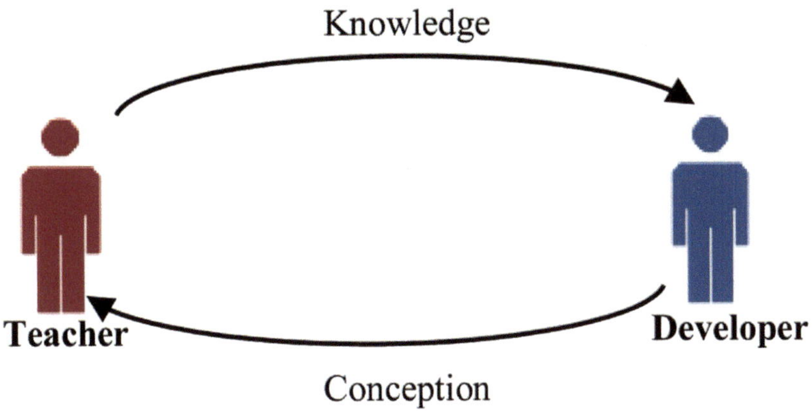

FIGURE 9.2 The teacher–developer relationship.

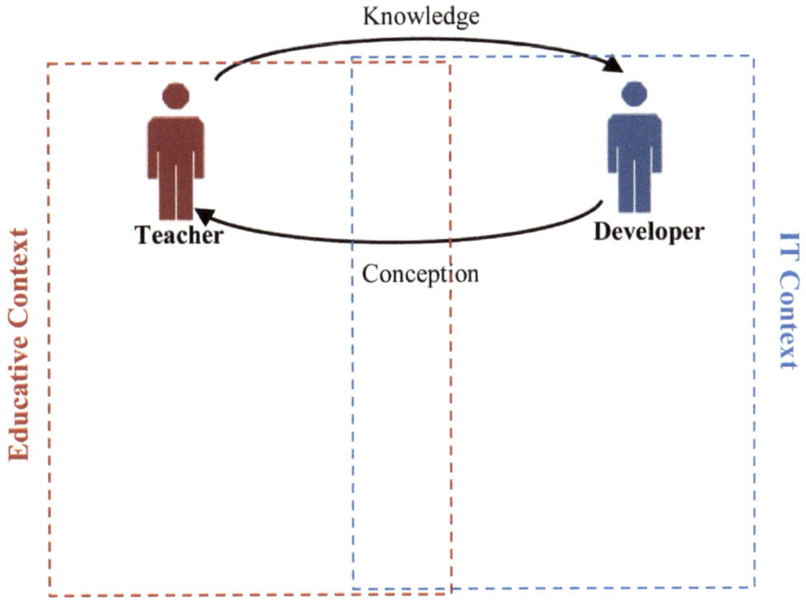

FIGURE 9.3 The extension of the relationship.

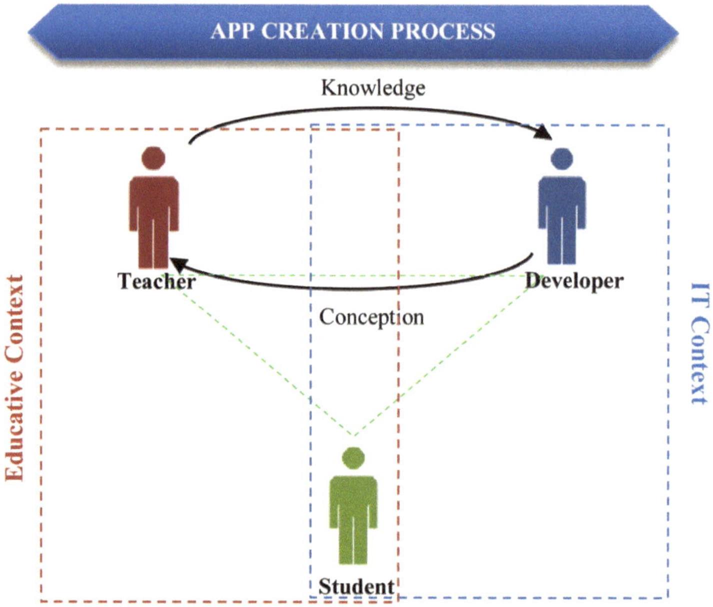

FIGURE 9.4 The student at the heart of the process.

FIGURE 13.2 Premium brands at social media

Source: Adapted with permission from www.pinterest.com.

FIGURE 13.3 Most preferred social platforms

Source: Adapted with permission from www.dbcinteractive.com.

Examples: Brands Using Social Platform

****TOPSHOP**: A blasting fashion brand used "Twitter" during London Fashion Week 2015 to approach youths for promoting the brand with exciting colors, shapes, and glamor. The brand engaged more than 3.8 million customers in its campaign named "LIVETRENDS." Sales were increased by 75% on various featured products.

****Dior J'Adore**: A dominant fragrance brand used "YouTube" for its 90-s ad commercial to get quick viewers buzz. The brand is known for its fabulous fragrance, glamorous femininity and sophistication. This footstep caused 1.1% increase in market share, 17% increase in sales, boosted brand name throughout the world.

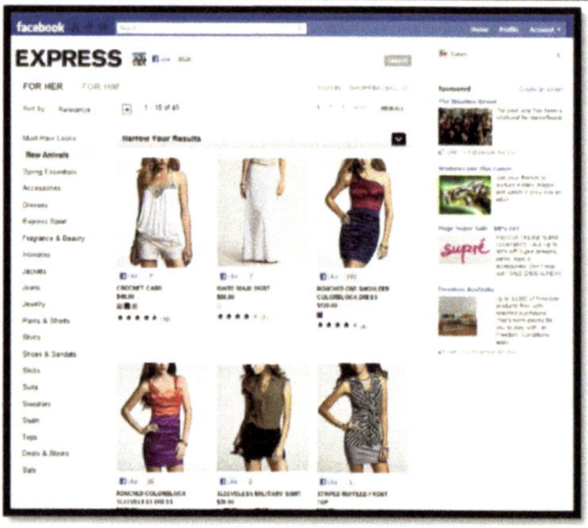

Women's and men's clothing at <express.com>.

FIGURE 13.4 Why social media is important?

Source: Adapted with permission from www.jeffbullas.com.

FIGURE 13.5 Specialties of Adova Soft.

Source: Adapted with permission from adovasoft.com.

FIGURE 13.6 Potential reasons for using social platforms.

CHAPTER 10

Integrating Social Media in Lesotho's Distance Learning: Gaps and Opportunities

SELLOANE PITIKOE

ABSTRACT

Technology has undeniably made a remarkable contribution in the social, academic, and professional spheres through digital learning. This chapter focuses on the contribution of social media in distance learning to widen learning opportunities that befit the 21st century. The chapter follows a literature review approach to address the following two research objectives: (1) it examines the current distance learning provision in Lesotho and its contribution to wider education access; and (2) it explores the possibility of integrating social media in Lesotho's distance learning provision. Although Lesotho piloted digital learning initiatives in the past to inform the integration of technology in the current system, those initiatives fell short due to lack of financial support. Arguably, Lesotho's current distance learning provision is predominantly a dual mode that comprises the traditional face-to-face and tutor–learner separation in space and time and uses the print media for instruction. The findings reveal the benefits of social media in distance learning provision and the economic value, where the learners can learn at a time and place that are convenient to them. The chapter suggests how Lesotho can improve the current distance learning from a print-based mode into a technology-driven provision that allows for collaborative learning and instant feedback.

10.1 INTRODUCTION

Incorporation of Information Communication Technology (ICT) in the different forms of education provision (formal, non-formal, and informal) is gaining a significant global momentum. ICT is an all-encompassing term that incorporates different forms of digital tools and platforms—hardware or software that are used for both communication and of social services such as distance learning.[27] ICT Scholars who advocate for integration of ICT in education applaud ICT for transforming learning into an attractive and equally accessible social service to everybody anywhere, while also enabling proficiency in digital skills that are much needed in the 21st century's labor market.[8] Another important web-based learning platform is e-learning, which is inclusive of all forms of networked computer-based instruction that is offered at a distance.[11]

The tremendous development in the usage of technological devices in our daily lives has made a remarkable contribution to social life through informal digital learning.[4,10] In earlier times, the technological devices came at unaffordable prices, which disadvantaged most of the intended consumers. The immobile nature of the digital tools confined the tools to one station, which made them less accessible to the socioeconomical incapacitated groups of society particularly in developing countries such as Lesotho. In the process, instead of achieving the intended mandate of facilitating digital learning, they instead escalated the digital divide—(the socioeconomic inequalities) between the haves and the have-nots.[11]

Lesotho is one of the smallest constitutional monarchy in southern Africa with an approximated area of over 30,000 square km with 2 million inhabitants.[26] Reportedly, the gender ratio in Lesotho is "97 males to 100 females." 70% of the population is rural based and highly affected by poverty and inaccessible social services including education. The increasingly deteriorating economy of Lesotho has resulted in Lesotho being categorized in the low-economy countries. Reports on Lesotho's education allude to the unique female literacy rates as compared with other sub-Saharan countries. For instance, the global reports by UNESCO and World Atlas reveal 84.9% of literate females against 67.7% of literate males.[24,29]

Like other developing countries, some of the challenges in the education access in Lesotho are the provision of equal opportunities to relevant education that addresses the employment needs of both the young and adult Basotho for poverty alleviation. Equally important is the poor and underdeveloped ICT infrastructure, which limits a wider access to the digital community engagement. However, in 2005, the country adopted a national ICT policy, which committed to nation-wide access to ICT tools in various forms.[15] Some of the intended strategic directions stated in the policy include: a need for national ICT literacy and training programs; promotion of technology-based distance learning to widen education access within the available limited resources. Despite ICT being mentioned in Lesotho's national education strategic plan, Lesotho does not have a policy that fully commits and directs an effective incorporation of ICT in the mainstream education.[18]

The progress made in Lesotho on technology-mediated learning uncovers that there were a number of ICT projects that were piloted within the mainstream education system, which never lived to see the light of the day. For instance, the New Partnership for Africa's Development (NEPAD) eSchools demo project, which was also running in other African countries other than Lesotho. The primary goal of NEPAD was to empower the primary and secondary school African youth with ICT skills while also promoting e-learning and internet use.[9] Another initiative was the establishment of a registered non-governmental organization (NGO) called SchoolNet Lesotho, a donor-funded initiative under the auspices of the National University of Lesotho (NUL) that is one of Lesotho's oldest institutions of higher learning directly funded by the government. NUL also piloted the Technology Enhanced Learning Initiative of Southern Africa through a telecenter that was based at the Institute of Extra-Mural Studies (IEMS) of NUL, although it is no longer in operation.[9] One of NUL's hiccups that prevent effective implementation of mobile learning is the underdeveloped ICT infrastructure, which impedes support for effective integration of mobile teaching and learning programs in Lesotho.[9]

To this day, there are three key players in the field of distance learning in Lesotho namely: the Lesotho Distance Teaching Center (LDTC), the Institute of Extra Mural Studies (IEMS), and the Lesotho College of education (LCE). Literature reveals that LDTC was established in 1984 as a functional department that is run by the Ministry of Education and Training (MoET).[14] LDTC's mandate remains a complementary function of the national mainstream provision that reaches out to the marginalized members of society such as the herd boys.[14] Initially, the department was implemented to offer correspondence learning activities to those learners who could not afford to fully participate in formal education.[12,13,14] It seeks to achieve four main objectives:[14]

- First, it helps the private candidates to complete the curriculum and sit for the examinations.
- Secondly, it provides basic education that addresses the diverse labor market needs.
- Thirdly, LDTC seeks to provide basic numeracy and literacy to out of school youth.
- Finally, it builds the capacity of other emergent NGOs that aspire to provide non-formal education (NFE) programs.

The main modes of content delivery were the radio programs, print material, and the weekend face-to-face classes.[13,14] One of the drawbacks that affect effective implementation of NFE teaching and learning programs in the developing countries is lack of trained human capital that can effectively drive the process forward. Equally important is the fact that NFE is accorded a lower status to that of formal education, hence suffers financial support and political will from the government (as is the case in Lesotho).[13,25] This implies an urgent need to explore the possibility of integrating social media technologies that are user friendly and accessible to most members of the society. It has been affirmed that the accessibility of cellular phones even to the most marginalized community of Basotho herders is a must; and they in fact indicated that they were already learning informally about HIV prevention using the radio application that came with their cellular phones.[19]

IEMS was established as an arm of the NUL to take the university out of the traditional classroom confines to the people by offering teaching and learning programs that addressed the felt needs of the people.[12,13] IEMS mainly targets the adult members of the community, who had missed out to enroll in institutions of higher learning such as NUL. The institution offers part-time programs in both formal and NFE using print and face-to-face learning modes.[12] Initially, the programs covered Bachelors' degree before the recent introduction of a Masters' degree level. Although IEMS made a significant impact in providing market-related skills to the learners, the personal experience of the author was that the face-to-face mode coincided with the job responsibilities of most learners. Equally important is the travel and accommodation costs that the learners incurred while attending the class meetings and the peer classroom working groups.

Based on these challenges, a need for social media devices in IEMS would not only address these challenges but it will also attract more learnership because of ease of access beyond the time and place confinement.

The LCE-offered Distance Teacher Education Program (DTEP) was started in 2001 under the former National Teacher Training College (NTTC), which was later renamed LCE.[16] In view of the rationale behind this strategic direction, emphasis was on the role played by distance learning as a second opportunity to those learners, who could not enroll in institutions of higher learning due to various reasons. Therefore, the program intended to develop the competencies of under-qualified primary school teachers for effective productivity under minimal job disruptions. This category comprises those teachers, who hold a high school completion certificate and a minimum of 2 years teaching practice.[16] The fact that such learners study autonomously with limited face-to-face contact with their tutors calls for a need to capacitate them on the nitty-gritties of distance and technology-based independent learning upon enrollment. However, evidence from Lesotho reveals that such learners did not receive prior preparation on the expectations of the journey that they were embarking on.[14,16] What is also lacking in the findings is the contribution by ICT to the successful implementation of DTEP.[13]

While it is commendable that Lesotho has higher adult literacy rates, care must be taken on the quality of education as well as the gender disparities surrounding education access. Some of the key impediments to education access include the herding practice that seems to override the importance of education in Basotho culture.[20] For instance, herding for males is considered a rite of passage and a definer of manhood in Basotho culture. As a result, the nomadic nature and the rugged topography of the herding areas limit access to education by males. Additionally, those herders who enroll in herders' NFE programs are forced to drop out of the system due to the nomadic nature of their career as well as the limited coverage of the herdboys' NFE learning posts.[21] Therefore, while boys chase animals during the day, their female counterparts enroll in formal education.

Where there is a dire need to improve the quality and access to education at any given time, the e-leaning-based technologies/platforms become an effective remedial action.[9,25] Emphasis though is on the need for the educators to be equally ICT competent to appropriately apply the new technologies in their teaching and learning practices. This notion resonates with Lesotho's situation in that, given the underdeveloped nature of the

ICT infrastructure, it can be argued that the teacher's competency levels are also not at par with the 21st century's demands and standards. Therefore, incorporating social media in distance learning programs of Lesotho may seem quite remote.

This chapter explores the potential contribution of social media in distance education provision to inform Lesotho's distance education policy and practice on how to design programs that befit the millennial development standards. The author intends to challenge Lesotho's scholarship for further research in the field of distance education and social media broadly and come up with relevant case studies that Lesotho can adopt and customize to "her" needs to address the existing gender disparities in education access. There are two research objectives addressed in this chapter: (1) it examines the current distance education provision in Lesotho and its contribution in widening access to education and (2) the chapter explores the possibility of integrating social media in Lesotho's distance learning provision. The discussions generated from this chapter will help Lesotho's distance learning policy and practice improvement.

10.2 LITERATURE REVIEW

Use of technology in the 21st century unprecedentedly increased the use of mobile media devices such as cellular phones, which came in smaller sizes at an affordable cost, hence reducing the past economic gap that hindered wider access to computers. Set against the backdrop of distance education provision in Lesotho, this chapter interrogates the potential contribution of social media in distance education programs offered in Lesotho. Social media includes all available digital applications mobile devices such as smart cellular phones, tables, and iPads. All technology-mediated learning activities, where mobile media devices are involved, can be grouped as social learning due to the interaction that takes place through peer learning activities.[1] The need to enhance a collaborative learning necessitates the incorporation of social media tools in teaching and learning activities.

10.2.1 SOCIAL MEDIA

The rigorous evolution of social media tools has therefore resulted in different definitions of social media by different scholars. In this chapter, the author adopted the definition of social media from Vaičiūnienė et al.[27] as a combination of various forms of digital devices, which may include but are not limited to: "collaborative projects, blogs and microblogs, content communities, social networking sites, virtual game worlds, and virtual communities. Various technologies list blogs, picture-sharing, vlogs, wall-postings, email, instant messaging, music sharing, crowdsourcing, etc."

The portable nature of the devices makes them easy to carry around and use for learning purposes at any time and place convenient to the users. The critical role of social media in teaching and learning is the one that combines both the traditional teaching with technology-based learning techniques while also promoting collaborative learning.[27] The autonomous nature of social media also places learner-directed content development and information sharing at the fore.[8] Evidence highlights that computer-based learning develops social relationships among members of the digital communities and how those relationships are important for the meta-cognitive and social well-being of the learners through informal social club[s][23] that can later be tapped on as a learning resource. Similarly, interactive learning increases learner's retention and participation in the learning activities[30]. Although social media is used globally to facilitate teaching and learning activities, there exists a knowledge gap on how Lesotho uses social media in distance learning programs, the gap which this chapter seeks to address.

Larreamendy-Joerns et al.[11] define distance education as "just-in-time learning," where the learner and the tutor are separated by space and time. Therefore, tuition and tutorials largely rely on different forms of media (audio, television, print, and computer) to reach out to the massive and widely dispersed learners. The growing research evidence argues that while technology-based learning is not foreign to distance learning, yet the incorporation of social media in distance learning reinforces learner-directed learning.[5,8] Research also reveals that online chat as another form of social media that uses instant messaging system between two or more people also can make significant contribution to teaching and learning practice through its chatroom application. The instant message delivery and feedback compliment the "just-in-time-learning" character of distance learning.[7] Given the fact that online chats have also become popular in Lesotho, the tool can potentially benefit Lesotho's distance teaching and learning provision. This communication application can benefit both the learners and the distance learning tutors with posting of questions and real-time online feedback, which is currently lacking in Lesotho's provision.[7] Social media comes in a wide array of social media networks and tools, which mainly facilitate digital communication and collaboration.[3]

10.2.2 FORMS OF SOCIAL MEDIA

Social media networks refer to a digital society, where the like-minded people create and share information formally or informally for different purposes: be it work, education, or social life related.[3] Interaction among the members is largely technology based through the internet, personal profiles and pictures for personal identification, and interaction among the community members. These networks can be grouped into three categories as follows:[3]

- Online communities such as: WhatsApp, Facebook, Ning, MySpace, and Twitter;
- Content sharing and organizing sites, which include platforms such as Delicious, Digg, Flickr, YouTube, Dailymotion, and Really Simple Syndication (RSS); and
- Content creation and sharing websites, such as: Blogger, Google Docs, Wikipedia, and WordPress.

Each social media tool has a specific contribution in teaching and learning practices. Nonetheless, it must be noted that some overlaps may be experienced based on the multifunctional nature of that particular social network site. Some of the benefits that come with social media are discussed in this section.

10.2.3 BENEFITS OF SOCIAL MEDIA

The phenomenal technology development has intertwined digital applications' use in our daily life routine as evidenced by massive introduction of computers, mobile devices, and improved software as communication platforms.[3] A surmountable number of teachers and students predominantly subscribe to social media and internet use which attests the influence that social media has on teaching and learning.[3,28] Despite of its infancy stage in research, some of its benefits in teaching and learning activities are:

- Firstly, the technology upgrade from Web 1.0—"static web" to Web 2.0—"social web" has transformed the one-way traditional classroom type of education which closely resonates with Web 1.0 into an interactive two-way communication process.[2,9] While Web 1.0 could not provide an instant feedback, the interactive nature of Web 2.0 allows for timely feedback and information sharing.[3,28] Collaborative learning and consultations can be facilitated through the "consultation patterns."
- Secondly, social media enables effective communication between the learners and the teachers.[6] Teacher and learner participation in digital communities breaks the rigid classroom setup by opening communication lines broader and allowing the introverts an opportunity to freely express themselves in a conducive social environment.[6] Such communication could be facilitated through media platform such as Facebook, WhatsApp, and Twitter, etc.
- Thirdly, social media improves learner engagement in the teaching and learning activities. The package comes with some attractive features that appeal to most of the learners and the teachers.[3] Media platforms such as the education Google Apps can potentially be resourceful for accessing and sharing of important information at a lower cost than it would have been if they had to purchase textbooks.[3]
- Finally, social media reportedly cultivates social interaction and a sense of a community among the learners and their teachers while at the same time strengthening the autonomous learning among the learners.[3,6] In wrapping up the benefits of social media in education, the benefits of social media in education include:[6] less costs, isolation reduction, overcoming the cultural diversity and tolerance barriers, collaboration with the likeminded, widening access to education outside the classroom walls.

In summary, the animal nature of human beings makes them to function better in social settings. Therefore, the collaborative nature of social media platforms supports cooperative learning while also utilizing the consolidated thinking process, where each member contributes to problem-solving for the benefit of all. The generated ideas are further cemented by the social bond that the group members would have already established.

Although research reveals the undeniable positive effects of technological developments that have brought about a social change in the way people lived and learned, the developing countries such as Lesotho are lagging. Given the high number of unschooled male Basotho, social media platforms can serve a remedial action of widening access to informal education to address the out-of-school children, youth, and those adults, who could not attend formal education during their childhood. The profound impact of poverty as an impediment to most Basotho children to access education could significantly be counteracted by integrating social media in education as it comes at a much lower cost than the tuition cost of formal schooling. Young as social media may be, it has gained an impressive momentum in the field of education as evidenced by the best practice models from elsewhere in the world.

10.2.4 BEST PRACTICE PROGRAMS IN SOCIAL MEDIA

The viral usage of technology worldwide has made a significant impact on mainstream education in various forms. Interest is evidenced by the comfort that learners from institutions of higher learning portray in the dynamism that comes with technology and its positive effects on social learning.[1] This section focuses on some of the best practice models of distance learning provision that Lesotho can learn from and customize to meet the country's education needs. One feasibility study has revealed some positive results on the impact that social media networks have on teaching and learning in the health professional.[22] Over 15 weeks of online collaboration, the students were tasked to independently design an online strategy that would facilitate online collaborative learning for health interprofessional education programs. Upon completion of the timeframe, the students commented on the possibility of effectively implementing collaborative learning for multidisciplinary learners such as those, who enroll for distance learning programs. However, the results emphasized the need to exercise care in the program and content design.[22]

Other researchers believe that cell phones have the potential to increase the equality and education opportunities by removing barriers to effective learning. For instance, the University of Pretoria (UP) in South Africa (the only neighbor to Lesotho) incorporated social media in the teaching and learning activities by using the mobile devices. The approach maximized the short message system, which was used as a study guide and learner support system. The system was also used to administer quizzes, and mini-lectures on complex topics. Two years later, UP had successfully implemented a two-way communication system that allowed for text receipt and response functions between the students and the tutors. In its own assessment of the strategy, UP concluded that the approach did not only improve the academic performance of the students but it also bore a positive impact regarding the learners' course completion.[7] It can also be argued that the approach was economical in terms of money and time, while also attractive to the 21st century learners, who are technology savvy.

10.3 MATERIAL AND METHODS

This chapter followed a literature review methodology, which examined various literature sources in the form of reports and journal articles. These sources were analyzed for distance learning and social media. The keywords that guided the author's search were distance learning, social media, technology-mediated learning, online learning, media devices, ICT, and e-learning. A total of 30 sources were identified and included in the review process. The data analysis was done based on the relevance of the content to the two research objectives and the benefits of social media in promoting learning. This objective-based analysis later generated the key themes for the research findings. The author also searched for best practice models elsewhere in the world, where social media was integrated in distance learning provision. The outcomes from this chapter will inform Lesotho's distance learning policy development and practice while also adding to the existing knowledge base on distance learning and social media. Researchers from other countries, who have similar research interests, may also equally benefit from these findings.

10.4 RESULTS AND DISCUSSION

10.4.1 LESOTHO'S CURRENT DISTANCE LEARNING

The advent of technology has not only transformed the world into a global village, instead more institutions of higher learning around the world have incorporated the technology-mediated education both in traditional classes and in distance-teaching and learning programs. The undeniable increase in learner enrolment rates in technology-based distance learning programs is evidenced enough of the crucial role of digital tools in delivering and upscaling distance learning.[7] In some countries such as Lesotho, distance learning is perceived as a second-class type of education as opposed to the high recognition given to formal education. However, the author argues that given equal opportunities and learner support, both the distance and the formal education learners can attain similar learning outcomes and competencies.[7]

Much of this technology comes with digital applications that open access to a learner-centric approach that is accessible anywhere and anytime convenient to the learner. Central to this provision is social interactions with the digital communities of learning and sharing of learning resources and material. The portable nature of social media tools makes them economic in that the social media applications come free of charge and can instantly connect a diverse group of learners. The smaller size of the social media devices makes them easy to carry and easy to use under minimal supervision. The online chat system facilitates real-time communication and instant feedback to a large group of learners and the tutor simultaneously, which is the missing link in most traditional distance learning provisions.[7]

Distance learning in Lesotho is mainly offered by the IEMS—a division of the NUL, LDTC—a government department, and LCE—a teacher training college. Although the former caters for adult learners from all walks of life, who have missed out their opportunity to enroll in formal tertiary institutions of higher learning, the LCE offers programs for teachers, who have not gone through a teacher training course, however, LDTC compliments the formal provision by reaching out to the marginalized groups of society such as the herders. It was also discovered that the three institutions offer their programs through a dual mode—a combination of a face-to-face and an independent learning. The medium of instruction is largely print media. While the reliability of print as a written record is underscored, the medium limits interaction and instant communication and feedback between the learners and the tutor. The gap can be filled with social media tools because they allow for real-time communication.

The findings also revealed that the learners, who enroll in distance learning programs in Lesotho, are not capacitated in advance on the expectations of autonomous learning. As a result, they cling on to the "old school" expectations, where the face-to-face becomes the main mode of learner support; the absence of which demotivates learning and academic performance. It was also unveiled that over and above the dual mode,

LDTC also employed audiotechnology in the form of radio programs as a medium of instruction. Amidst the trending adoption of technology-mediated learning as a global outcry, Lesotho is still behind.

Some of the reasons behind the delayed adoption of digital tools in distance learning in Lesotho are: (1) The relegation of all forms of NFE to a lower status leads to lack of government commitment and financial support to such programs. Notably, learner support in distance learning programs forms the backbone of the provision. (2) Currently, the limited ICT resource base in Lesotho does not place the country in a good space for adequate provision of learner support. Therefore, more research is needed in learner support that would investigate the types of learner support, which Lesotho can offer and how the country can possibly implement such proposed learner support initiatives. (3) Similarly, the upscaling of distance learning in Lesotho is hampered by the limited literature on how to go about integrating social media in Lesotho's distance learning.

10.4.2 DISTANCE LEARNING THROUGH SOCIAL MEDIA

The author mentions that integration of social media in distance learning does not only align the provision to the current demands of the 21st century but also makes education more appealing and empowering in preparation for absorption in the labor market. The social media tools can significantly contribute new ways through which Lesotho can offer distance learning programs. Some of the functions that social media tools can play in Lesotho's current provision include: (1) content creation; (2) ease of access to new knowledge; and (3) information sharing and imparting knowledge.[17] However, the success of the social media tools leans heavily on the healthy policy environment and government support.[17] The author concluded the following potential benefits that Lesotho can reap from the social media-based distance learning programs:

- Social learning: Social media brings people together in real time where they collaborate, learn, and share resources as common interest groups/learning communities/communities of practice. Learning comes in different forms: formal, nonformal, and informally, where learning becomes a shared community activity for the benefit of all. Social media also facilitates the development of relationships within the digital society members and a social capital that members can tap into for peer support during their learning.

- Wider access to quality learning: Social media technology takes learning outside the traditional classroom to the people in that it is not time and place confined. Technology becomes the mode of delivery that simultaneously covers a widely spaced population of learners, who are separated by space and time. Similarly, the learners and the tutors can communicate instantly.
- Liberal learning: The digital nature of social media makes learning mobile and accessible while also broadening choices of the content to meet the diverse needs of the learners. Through the social media tools, the learners are at liberty to direct the learning process by choosing what they want to learn, when they want to learn, how they want the learning to be conducted, and when they intend to complete.
- Transformative learning: Social media transforms the traditional teaching and learning practice into an e-learning platform that comes with features and applications that are accessible online at affordable costs. For instance, members can download digital applications such as Facebook, WhatsApp, Twitter, and YouTube to connect with other learners/tutors and learners using online chat either in groups or individually. The tutor mainly assumes the role of a facilitator while also learning from the contributions made by the learners as opposed to the teacher-directed traditional classroom setting. While the traditional classroom heavily relies on the face-to-face approach, digital learning on the other hand uses Web 2.0 learning tools, which allow for interactive learning and complements the face-to-face approach.

Integration of social media in distance learning will not only make education accessible to the learners but the education will also be of good quality. Central to e-learning is learner collaboration, which nourishes the cognitive well-being of the learners while also empowering the learners to be accountable for their own learning. While distance learning may be relegated to a lower education status, Lesotho's policy and practice may have to reconsider the value that distance learning adds to education access by providing a second opportunity to those learners who could not access education due to space and time limitations. Therefore, through social media, opportunities for further learning become wider, diverse and learner centric. Education provision is a joint venture between the government, the civil society, the local governance, parents, and youth. This implies a need for national dialogues that would give the key stakeholders a voice on how to incorporate social media in the current distance learning provision in Lesotho.

10.5 LIMITATIONS AND CONCLUDING REMARKS

Indeed, the current distance learning provision in Lesotho suffers limited resources that can support and sustain distance learning. Equally important, the gender disparities in Lesotho's literacy rates, where more females have access to education than males, can potentially be remedied through social media tools to bridge the education divide. The paucity of a broader evidence-based research to integration of social media in distance learning will not only misinform Lesotho's policy and practice reform but it will also delay the progress of making the current provision more appealing to the learners.

While the benefits of social media in distance learning are unquestionable, some of the glaring challenges in transforming the current mode into digital learning are: (1) The country is currently going through political instability, which may potentially take a toll on the financial commitments of the country. (2) While the ICT policy is in full swing, Lesotho's ICT infrastructure remains fragile. Given the rugged topography of Lesotho, the reliability of the network in supporting social media in distance learning is questionable. (3) The ICT policy commits to national ICT trainings. However, progress in this area for the past 14 years is very thin on the ground. (4) While Lesotho also submits to the global outcry to include technology in the teaching and learning activities, the unhealthy economy of the country significantly delays the planning and implementation processes. Lesotho is classified under the low-income countries. Therefore, much of the infrastructural developments rely on external funding, which comes with donor-driven expectations that are mostly not contextualized to meet the local needs.

An effective implementation of social media in distance learning also requires both the tutors and the learners to be equally technology competent to drive the process forward. The poor ICT infrastructure leaves a lot to be desired on the readiness of Lesotho's tutors to teach the learners on how to navigate distance learning using the social media tools. This gap implies a need for ICT training for the tutors in such a way that they will be confident enough to pass their skills on to the distance learners.

Evidence-based research identifies some of the good practice models that Lesotho can tap into and adapt to the current provision. For instance, the pilot feasibility study unveils the significant contribution of collaborative learning in the health interprofessional learning program.[22] Another model that can be applicable to Lesotho comes from the neighboring South Africa. In this case, the UP incorporated the text messaging into the teaching and learning practices. The results of which were higher course completion rates and improved learner performance. Given the importance of collaboration in social learning activities, both models can be adopted and incorporated into the current practice. The UP model is particularly friendly to Lesotho's situation, where cost for infrastructural support is still a challenge.

Lesotho like her neighbor, South Africa, may want to explore the use of the mobile devices that seem to be popularly owned by most Basotho including the marginal groups of society such as Basotho herders. Most if not all of immediately these mobile devices come with the short messaging system application that is commonly used for communication among Basotho because it does not have installation costs. Equally important is the affordability of sending and receiving messages. Given the poor ICT infrastructure and the rugged topography, which may hamper communication, the short messaging system has survived the test of time and has become the most reliable mode of communication for Basotho herders, whose semi-nomadic career is mainly practiced in hard to reach areas of Lesotho.[19]

10.6 SUMMARY

This chapter examines how Lesotho offers distance learning programs in a bid to widen access to education beyond the confines of space and time in such a way that it can be easily accessible to the marginalized groups of the society. The chapter also explores further opportunities that Lesotho can tap into and integrate social media in its current distance learning provision to make the provision to be more appealing to the intended consumers. The findings presented in this chapter were reached through a review of literature in the following fields: distance learning, social media as well as social learning where a total of 31 literature sources were engaged into.

The 21st century has turned the world into a global village, where collaboration and communication are made easier by using the digital tools. Most of these tools come with applications and technology platforms that allow real-time communication and feedback provision. Some of the popular digital tools and platforms include the computers, mobile media devices such as cellular phones, tablets, iPads, iPhones; applications such as Facebook, Twitter, and WhatsApp. Literature categorizes the social media networks into three groups namely the online communities, the content sharing and organizing sites, and the content creation websites. It was also found out that the convenient nature of technology has made education accessible anywhere and anytime of the learners' choice. As a result, technology has gained a significant momentum in the field of education and research.

Lesotho like other countries elsewhere in the world has subscribed to the need for digital learning. For instance, despite her poor ICT infrastructure, Lesotho developed and adopted an ICT policy in 2005, whose intent was to widen access to ICT tools and technology nationwide while also supporting technology-mediated distance learning. However, not much progress has been reported in this area. This is evidenced by the current distance learning provision that still relies on print media offered largely on a face-to-face mode. On a positive note, it was also uncovered that most Basotho own mobile devices such as cellular phones so much that even the most marginal groups of herders own cellular phones. This can present a ray of hope, which Lesotho may want to explore for future integration if social media in distance learning.

Some of the reasons that can account to the delayed implementation of social learning include the limited government support to NFE activities because Lesotho does not have a ratified policy that supports the implementation of distance and NFE activities. Also, not much research has been done that can inform the policy and practice on the necessary change and how such change can be implemented. Currently, the limited ICT resource base in Lesotho does not place the country in a good space for adequate provision of learner-support.

Some of the identified benefits of including social media in education that Lesotho can potentially benefit are: firstly, social media brings people together in real time where they collaborate, learn, and share resources as common interest groups/learning communities/communities of practice. Secondly, social media technology takes learning outside the traditional classroom to the people in that it is not time and place confined. Thirdly, social media transforms learning into a liberal activity that is easily accessible with broader choices of the content to meet the diverse needs of the learners. Fourthly, social media transforms the traditional teaching and learning practice into an e-learning platform.

The chapter suggests that Lesotho may want to explore the feasibility of integrating collaborative learning in the current distance learning provision with emphasis on the use of mobile devices. Judging from the positive results that the University of Pretoria achieved in integrating the text message system for learner support, curriculum guidance as well as in the administration of mini lectures and quizzes, Lesotho may want to explore this model further for incorporation into the current distance learning provision.

KEYWORDS

- **distance learning**
- **e-learning**
- **Information Communication Technology**
- **mobile devices**
- **social learning**
- **social media**

REFERENCES

1. Blair, R.; Serafini, T. M. Integration of Education: Using Social Media Networks to Engage Students. *System. Cybernetics Inform.* **2014,** *12* (6), 28–31.
2. Cole, C. Social Media Best Practices for Nonprofit Organizations: A Guide, 2014, p 78. http://www.ccghr.ca/resources/social-media-modules (accessed June 1, 2018).
3. Faizi, R.; El Afia, A.; Chiheb, R. Exploring the Potential Benefits of Using Social Media in Education. *iJEP* **2013,** *3* (4), 50–53.
4. Galanis, N.; Mayol, E.; Marc Alier, M.; García-Peñalvo, F. J. Supporting, Evaluating and Validating Informal Learning. A Social Approach. *Comput. Human Behav.* **2016,** *55,* 596–603.

5. Gikas, J.; Grant, M. M. Mobile Computing Devices in Higher Education: Student Perspectives on Learning with Cellphones, Smartphones & Social Media. *Internet Higher Educ.* **2013,** *19*, 18–26.
6. Gurcan, H. I. Contribution of Social Media to the Students' Academic Development. *Int. J. Inf. Educ. Technol.* **2015,** *5* (12), 965–968.
7. Hanover Research. Distance Education Models and Best Practices, 2011, p 39. http://www.hanoverresearch.com (accessed June 26, 2018).
8. Hylén, J. Mobile Learning and Social Media in Adult Learning. Literature Review Prepared for the ET Working Group on Adult Learning. European Commission; Brussels, Belgium, 2015, p 15.
9. Isaacs, S. ICT in education in Lesotho—Survey of ICT and Education in Africa: Lesotho Country Report, 2007, p 10. http://www.infodev.org (accessed June 26, 2018).
10. Koşar. G. A Study of EFL Instructors' Perceptions of Blended Learning. *Procedia Soc. Behav. Sci.* **2016,** *232*, 736–744.
11. Larreamendy-Joerns, J.; Leinhardt G. Going the Distance with Online Education. *Rev. Educ. Res.* **2006,** *76* (4), 567–605.
12. Lefoka, M. M.; Panda, S. K. Financing Distance Education: Case Study of Lesotho Distance Teaching Centre, Southern Africa. *Middle East. Afr. J. Educ. Res.* **2012,** *1*, 19–40.
13. Lephoto, H. M. Transforming Higher Education in Lesotho Through Open and Distance Learning: Issues and Challenges. *Formamente* **2006,** *1* (3–4), 117–132.
14. Maieane, A.; Malefane, J. Overview of Lesotho Distance Teaching Centre's Distance Education Model. In: *Learner Support Services: Case Studies of Deasa Member Institutions*; Nonyongo, E. P., Ngengebule, A. T., Eds.; University of South Africa: Pretoria, South Africa, 1998, p 15.
15. Ministry of Communications, Science and Technology. Lesotho ICT Policy, 2005, p 45. http://www.lesotho.gov.ls (accessed Sept 16, 2016).
16. Mohono-Mahlatsi, L.; van Tonder, F. The Effectiveness of Mentoring in the Distance Teacher Education Programme at the Lesotho College of Education: Student Teachers' and Tutors' Perceptions. *South Afr. J. Educ.* **2006,** *26* (3), 383–396.
17. Möller, C. Social Media and Journalism (Chapter 1). In *From Traditional to Online Media: Best Practices and Perspectives;* Stone, M, I. Dohel, L., Eds.; 14th Central Asia Media Conference; The Representative on Freedom of the Media Organization for Security and Co-operation in Europe, Vienna, Austria, 2013, p 7.
18. MoET. Unpublished Lesotho Education Sector Strategic Plan 2005-2015.Maseru-Lesotho, South Africa, 2005 p 110.
19. Pitikoe, S. Male Herders in Lesotho: Life Histories, Identities and Educational Ambitions. Ph.D. Thesis, University of KwaZulu-Natal, Pietermaritzburg, South Africa, 2016, p 354.
20. Pitikoe, S. Basotho Herders Learn Through Culture and Social Interaction. *Learn. Cult. Soc. Interact.* **2017,** *13*, 104–112.
21. Pitikoe, S.; Preece, J. Herder Identity in Lesotho: Implications for Non-formal Education. *Int J. Lifelong Educ.* **2016,** *35* (6), 607–628.
22. Pittenger, A. L. Instructional Design and Assessment: The Use of Social Networking to Improve the Quality of Interprofessional Education. *Am. J. Pharm. Educ.* **2013,** *77* (8), 1–9.

23. Prins, E.; Toso, B. W.; Schafft, K. A. It Feels Like a Little Family to Me: Social Interaction and Support Among Women in Adult Education and Family Literacy. *Adult Educ. Q.* **2009,** *59* (4), 335–352.
24. UNESCO. Global Partnership for Girls' and Women's Education—One Year On, 2011, pp 2. http://www.unesco.org/new/en/education/ (accessed Sept 16, 2016).
25. UNESCO. Open and Distance Learning. Trends, Policy and Strategy Considerations. UNESCO: Paris, France, 2002, p 91.
26. UNDP. *Lesotho National Human Development Report: Leveraging the Power of Youth to Promote Human Development.* United Nations Development Program: New York, 2015, p 232.
27. Vaičiūnienė, V.; Romeris, M.; Mažeikienė, V. Social Media in Adult Education: Insights Gained from Grundtvig Learning Partnership Project "Institutional Strategies Targeting the Uptake of Social Networking in Adult Education (ISTUS)". *Soc. Technol.* **2012,** *2* (2), 473–482.
28. Vervaart, P. Role of Social Media and the Internet in Education. *J. Int. Fed. Clin. Chem. Lab.y Med. 23* (2), 1–4.
29. WorldAtlas. https://knoema.com/atlas/Lesotho/topics/Education/Literacy/Adult-literacy-rate. (accessed Sept 16, 2016).
30. Yeung, F. P. F.; Fallucca, A. Systems for Documenting Student Experiences and Outcomes (Chapter 1); In *New Directions for Institutional Research*; Wiley Periodicals, Inc.: New York, 2018, p 13. Published Online: 2018. http://www.wileyonlinelibrary.com (accessed June 1, 2018).

CHAPTER 11

Role of Social Networks in Healthcare for Developing Countries

JANA SHAFI, AMTUL WAHEED, and SUGAPRIYA DHANASEKARAN

ABSTRACT

This chapter focuses on issues, awareness, importance, demands, trends, and promising future in the field of health with a strong support of social media through mobile phones. Authors also emphasize that young generation of developing countries are facing problems in terms of resources, effective social networks, operating skills, education, financial crisis, and limited healthcare provisions. With the expansion of social media, it is a subject of research, data collection, and observation and a great support to other areas, which are fully or partially dependent upon it. It is necessary for leaders in the developing nations to understand and promote its use.

11.1 INTRODUCTION

Digital social media is an imaginative convergence of digital knowledge and is a big industry for human behaviors, science, arts, communication, education, and social relations.[12] The different sorts of digital routes improve communications, education, and research by replacing conventional discourses to multidirectional discourses, which are not restricted to mobile and public media, social research, and customer relationship management (CRM). The incitement, which encloses impending of digital social media in community health, is clear. Though, the expert commentary or existing research forms a fascinating dispute for providing money and time in use of digitalization routes in open health sectors. It is not sufficient to mention stagnant case reports that use social media such as

Twitter, Facebook, and WhatsApp for public health issues, campaigns, and programmers to discuss the fitness issues; and more people have knowledge and understanding of digital media use that guides the way to take social media to the subsequent phase.

In this chapter, authors focused on major styles for constructing on the future holds of opportunities in public health sectors. Digital channels can make behavior change and influence Twitter, Facebook, and WhatsApp as an additional room of consumer check portal. Big Data's influence in the health area also pays attention to digitalize media to update approaches and interactions. Furthermore, studies in this chapter lead to the prospect of community healthcare's commitment and health issues with subsequent age group of digital medium exploitation as well as recommend a structure used for scheduling and estimating digitalize channel use.

11.2 SOCIAL MEDIA USAGE IN HEALTH SECTORS: CURRENT STATUS

Worldwide, there are 2.13 billion monthly active profiles (2017) on social media that include Facebook,[33] and YouTube being the global second rank search engine.[78] The consequence of comparatively deprived health understanding of cultural underground groups among various parts of the world has been mentioned in the concluding information of the Health Commission on Social Determinants.[66] An effective significant support in the job of developing racial alternative fitness issues is an elevated usage of digital medium between cultural alternative groups in several developing nations. Table 11.1 indicates the usage of mobile applications (apps) related to medical issues and healthcare activities in urbanized nations. Univision's Hispanic Patient Journey Case Study showed that 57% of Hispanic families and their friends are main basis for data among the Spanish communication persons in the United States, compared to 41% non-Hispanics (41%).[74]

In addition, a few other studies in the United States are affirmative for under communities of the people's usage communal medium policies that include Facebook and Twitter, likewise to nonminority communities.[36] Furthermore, smartphone usage is lower between underground inhabitants, who use highly sophisticated applications on their cell phones to connect online.[22]

TABLE 11.1 Mobile Apps Related to Medical Issues and Healthcare Activities.

Information management		Consulting and communication		Information gathering and reference		Monitoring and patient management		Training and medical education	
Application name	**Services**	**Application name**	**Services**	**Application name**	**Services**	**Application name**	**Services**	**Application name**	**Services**
Evernote	Note-taking and organization	Doximity	Social networking site for MDs	Epocrates	Health and medicine orientation	Diagnosaurus	Differential diagnosis	Med Page Today	Healthcare instruction
Notability	Note-taking and organization	–	–	Dynamed	Health and medicine orientation	Pocket Lab Values	Research laboratory position	Quantia MD	Healthcare instruction
iAnnotate	PDF viewer	–	–	Skyscape/ Omnio	Health and medicine orientation	Lab Pro Values	Test center location	–	–
Good Reader	PDF viewer	–	–	Micromedex	Medicine orientation	Archimedes	Medicinal calculator	–	–
Box	Cloud storage and file sharing	–	–	Dynamed	Health orientation	MedCalc	Medicinal calculator	–	–
Dropbox	Cloud storage and file sharing	–	–	UpToDate	Health orientation	Mediquations	Medicinal calculator	–	–
Google Drive	Cloud storage and file sharing	–	–	Medscape	Health orientation	Calculate	Medicinal calculator	–	–
–	–	–	–	Johns Hopkins Antibiotic Guide	Health orientation	AHRQ ePSS	Showing and prevention tool	–	–
–	–	–	–	Sanford Guide to Antimicrobial Therapy	Health orientation	–	–	–	–
–	–	–	–	Medpage Today	Health bulletin	–	–	–	–

Recent study in the United States by ModermMom.com and Fleishman–Hillard indicated 63% of mothers' online assessment and study toward significant purchase of their final selections. Most mothers (>85%) visited social networks such as Facebook and Twitter regularly,[79] making them one of the rapidly rising demographics online. Ethnic minorities from the United Kingdom mentioned that they have higher levels of internet access through mobile phones than the joint accounts.[52] India is being converted into the biggest marketplace for Facebook in 2019 through mobile phones (online access) with >250 million consumers with 200,000 users being added every day. In Japan, people are spending more than 2.9 h per day online, where Twitter pages are bigger and are used more than Facebook. Sweden is the most linked country with social media globally, whereas Brazilians have the higher Google+ user's population. Russia maintains >50 million blogs, whereas China has a leading communal network named "Qzone" with >530 million users.[75] On behalf of wellbeing-related issues, these global contribution outlines are of utmost importance.

The most energetic involvement is contribution in digital networks, which are difficult to spread to ethnic minorities, new defenseless persons. New mothers can play a significant role in spreading awareness of healthcare and how to avoid ill-health through their social media and digital networks. The capability to influence these digital networks involving this approach for healthiness eventually meant that fitness condition communication (message) is used to develop into further prominent plus important communal throughout the reliable speeches of neighbors, associates, colleagues, households, and other effectors. Furthermore, communication can be made at a very fast pace through use of digital media. Facebook post of the Marketing charts can reach about half of its customers in the first 30 min after posting.[61]

In case of viral distribution of an electronic mail, 23.95% of viral e-mails were unlocked within the initial (primary) time after delivery,[77] which clearly illustrates that digital channels (such as E-mail, Facebook, and other social media) also have distinctive capability to accomplish fast distribution of significant information. In this chapter, authors focus on different paths in which digital media and health technologies can respond to country's fitness care reaction, at individual, groups, and regional levels. The chapter also describes important information of existing body of knowledge at present conditions, and instant potentials for digital networks in the fitness sector.

11.3 HEALTH SYSTEMS IN DEVELOPING NATIONS

Emerging nations suffer an escalating rate in noncommunicable long-lasting diseases; also infectious diseases pose a continuous risk. Popular health issues in developing nations are hypertension, obesity, heart disease, and diabetes.[71] The mutual effect of chronic, communicable, or noncommunicable illnesses is expressed as "double load" aimed at emerging nations.[11] Thriving pains to decrease the double load of sickness drives to improve quality of life for thousands of patients. Mobile-health technology deals with kind of expectation on equal sides. The foremost avoidable reasons of noncommunicable infections worldwide are: tobacco, malnutrition, and little exercise. These are applicable even to heart sickness, diabetes, lung sickness, and cancer.[17,53] These are expected to increase in the developing world.[2,29,37,53,57,72] There must be unique care strategies for chronic, noncommunicable diseases, which may be unreachable in developing nations but is possible with the use of mobile phones.[3] Mobile phones have benefitted us in various ways, such as:

- The extended expectancy time for long-lasting diseases frequently needs initial, comprehensive grounded society fitness involvement.
- Decreasing enduring sickness frequently needs declining performances related by superior capital such as: regimes extraordinary in sugar plus fat, tobacco, less exercises.
- Chronic diseases conduct usually practices through numerous mechanisms of healthcare structure. This involves expert fitness specialists and synchronized unremitting healthcare.
- Prolonged medication need for long-lasting diseases, familiarizing subjects of admittance, fee, also quality of pharmaceuticals and devotion to treatment routines.

Countless factors limit performance of health systems in developing nations. Constraint on infrastructure and hospital possessions are strong in urban areas.[71] Burden of diseases, rate of disease, its impact on lifestyle of people and economic efficiency are immense.[45,46] Healthcare volunteers[63] are hard to hire and retain, particularly in remote areas.[5,60] Administrative and executive schemes are frequently absent or else feeble. On analysis of these constraints acknowledged by many areas, healthcare via mobile can be a big help in removing physical obstacles, maintenance, and provision for refining of feeble healthiness structure and poor communications.

11.4 IMPORTANCE OF MOBILE AND SOCIAL MEDIA IN HEALTHCARE

Healthcare affordability, access, and quality are major obstacles. Digital technology provides the direction of support by means of these tasks. There are possibilities of enhancing healthcare sector, such as: cellular technology-applied sensors, healthcare requests, remote and medicinal machines, patient products, etc. All digital expertise can lessen cost of healthcare through simplifying the provision of healthcare and relating individuals to their fitness care suppliers. Requests permit healthcare providers and patients to have access to medical records, tracking medications, laboratory analysis, evading medication alteration, guidance, augmenting portfolio control and dipping charges, monitoring compliance violations, tracking of infant growth at each phase of the prenatal period, helps to avoid miscommunication and confusions, recover patient's healthcare through supplying texts, increases focus on patient care and documentation, enhances responsiveness and patient satisfaction, and reference materials through mobile phones.

Multifaceted cellular phones healthcare apps can provide support in different zones, for example, management of chronic disease, awareness about communicable disease, monitoring of critical health indicators and training of healthcare workers. They allow access to software tools like physician or hospital locators, prescription reminders, medical references, calorie counters, and appointment notices. These mobile applications enable healthcare suppliers to speak on medical subjects, minute examination in addition to cure management, and any issue of the locality of patient or healthcare supplier.

11.4.1 ADVANCE APPLICATIONS IN MOBILE HEALTHCARE

iWander applications are aimed at Android O.S mobile phones, which are used for Alzheimer's disease or dementia patients to track patient locality through GPS utility of multimedia phones. If the person starts travelling from the house or other known locality, iWander app triggers an indicator gesture to the domestic member or warden to track their move. Based on the geographical axis, the dementia patient can be identified and safely brought home.[15] Digital media serve patients by aiding in emergency appointments.

A website known as TuDiabetes is used approximately by 500 diabetes-related patients for description and related occurrence and practice with gender, age, hypoglycemic measures, insulin pumps' usage, and healthcare-related complications. These applications will help, since distribution of familiar audience might make out other practices further paving the way to survive by specific fitness traumas. An early report showed an average person experienced six insulin-related health issues.[13] One major drawback is complication in connecting various devices. These inter-operational disputes can begin in numerous ways: (1) info devices unable to connect with each other; (2) here are disagreements in context of semantics, facts, also sharing files procedures; (3) assigning the wireless ways on a cloud method facilitates to connect subjects; and (4) constructs simplified info systems.

eMix is known as "Electronic Medical Information Exchange" of a cloud-based system to allow healthcare providers and patients to fetch their case reports. Individuals observe medicinal background, pathology reports, and therapeutic imaging reports as a safe delivery method, which facilitates the connection of a society to the medical data, where they are presently located. Similarly, another system program is known as 2net Platform established by Qualcomm Life. 2net Platform stores, transfers, supports, transforms, and shows automated medicinal instrument facts. Cloud System program is planned through various types of application and medicinal plans.[41] Healthcare providers and patients have access to their information on time that is an advantage in case of health crisis. Due to expansion of countless new cellular applications, it is vital to identify the one that are most caring and efficient.

iMedical Applications is the foremost channel for medicinal management. The forecaster suggests assessments and explanations online as medicinal expertise. Viewers can find out, which of the apps can yield the maximum optimistic analyses or which ones cannot. It provides references on the healthcare apps that are available or accessible in the market. Mobile Apps discovery is not an easy job. It takes huge time, aptitude, technologies, capital and skill to set up supportive mobile health apps. Mobile surroundings must promote and make possible its operations.

11.5 PROMISING APPROACHES FOR E-HEALTH IN DEVELOPING COUNTRIES

In present world, health information on mobile, wireless, and informative data communications is defined as Mobile-health (M-health). Mobile health interactions are classified based on wide geographic zones, local society, in-person patients, and suppliers. Two competent assignments signify small officer finding evaluation of M-health in developing countries.[9,24,39,40,42]

Through usage of SMS and calls or script messages, Krishna and his team[40] have studied and reviewed healthcare for 12 experimental regions, set-up for major developments in observance through medication intake, HbA1C, stress levels, asthma symptoms, rate of quit-smoking, and own effectiveness. Practice development was declared in minor unsuccessful programs, more fast diagnosis and demonstrating, and enhanced supervising and instructing. This research is being done in wealthier nations except in China.[40] A review statement in 2006 declared that today there is practically no literature on mobile technology as healthcare interference for HIV, malaria, tuberculosis, as well lingering circumstances in emerging nations.[39] Mobile interaction systems offer devoted and unique chances for mobile-health in wide topographical zones. Definite apps of M-healthdescribed in Figure 11.1.

11.5.1 ELECTRONIC MAIL

Listserv software controls the public subscriber's mailing lists to distribute information to patients or suppliers of clinical surroundings from web. While hopeful profits grip for e-mail interactions among doctors and their patients in context of lesser costs of healthcare even as upholding or improving the quality of disease administration and health advertising. Use of e-mail by doctors is still very limited e-mail with patients and lags the readiness of patients to interact with their doctors via e-mail. There is also an inconsistency among doctors' eagerness and real practice of e-mail interaction.

Number of factors may clarify these differences: The doctors vary in their skills and approach toward information technology; few may not be persuaded that patients welcome, require, and can talk by e-mail with their physicians; others are still coming up for healthy confirmation on service presentation and efficiency in adding up to patient approval and result that support such practice; and many are hesitant to do so because of alleged obstacles. The necessary Governmental support and orders for e-mail announcement to shift forward is iterated, and opportunities for prospect research are look upon out.

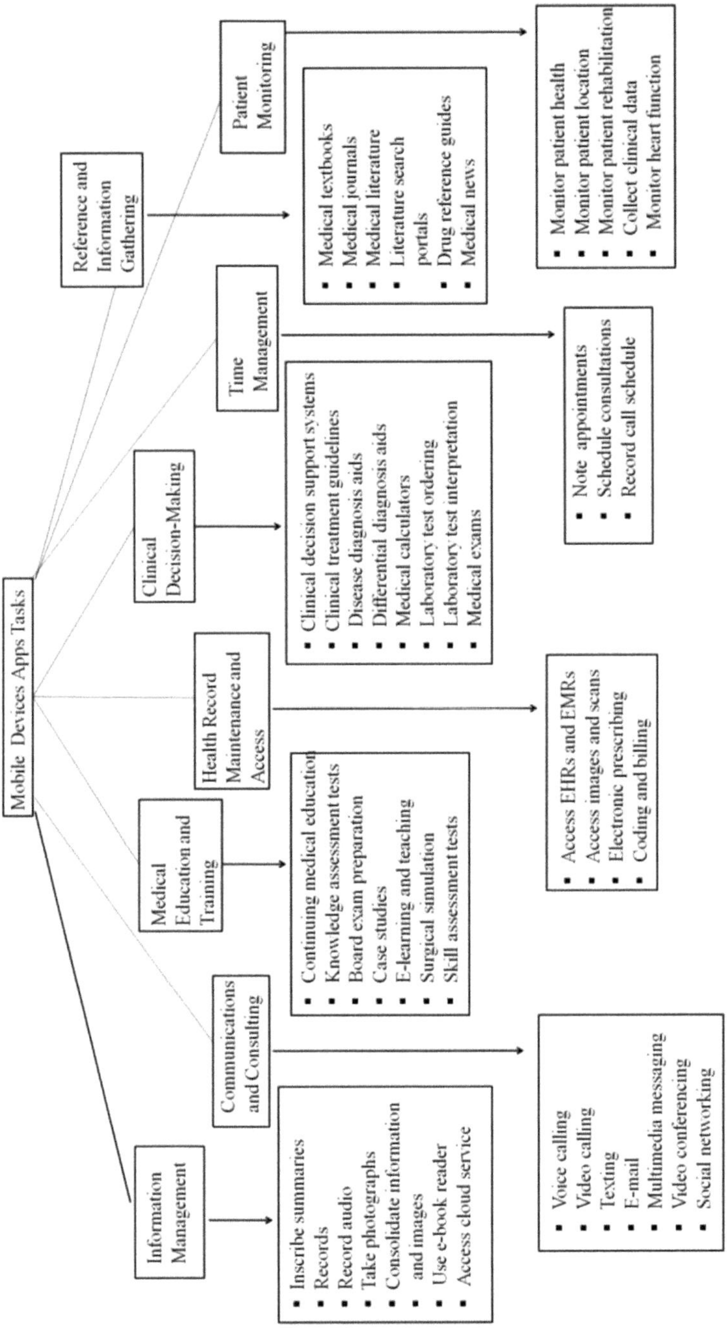

FIGURE 11.1 Mobile devices apps and tasks.

11.5.2 SOCIAL NETWORKING

Electronically connected groups of persons are shown with the help of social networking forms, which include several tools, in which the most distinctive and accessible is mobile text messaging. This facilitates to reduce disasters in varied geographical areas; for instance, a storm message alerts to the mobiles to save life of millions of people (fake alarms are always a big risk). Social networking can also be powerful in advertising health, for example, hopeful healthy activities by messages dispersed via text messaging. This device is also useful in carrying out harmful evaluations; arranges relief efforts; and get support of simultaneous persons and inform them to safe shelter, clean water, and healthcare. Personal digital assistants (PDAs) are used to put into a table and report details of survey and observation data. Mobile Text Posts are directed in native spoken message, to receiver by the Helpline of National AIDS numbers in Project Masiluleke in South Africa.[24] Mobile Text Messages are targeted to avoid practice of tobacco, smoking, enhanced nutritional choices, prevention of dangerous sex activities, violent behavior, and wound avoidance.

11.5.3 WEB SURFING

Internet Web surfing allows patients to avail medical data and medical surroundings, which simplify rapid diagnosis and get better communications among patients and health providers. Shortcoming of expanded availability of medical data information devoid of any value leads to incorrectness or to misdiagnosis. M-health upgrades the synchronized supply series, replacing a provider and only one distribution system with numerous dealers for varied logistical requirements for supply.

Web knowledge is an instrument to deliver numerous instructors and amenities, for allowing decentralized guidance. Restricted choices for superior training express patient interactions and are short of real-time messages. Web-based data entry is providing centralize online database and M-health operate as tool to reach and get back data from it. One of the studies recommends that basic method is to watch for malaria epidemics in East Africa, which might be economical.[41] Shortcoming of M-health directs toward risks of misleading information that directs to fake acts on disease epidemics or the unprofessional conduct of patients.

11.5.4 ELECTRONIC EDUCATION

Electronic education in health sector is with us from the day internet started. It had been 20 years that we can plan for the prospects of electronic health education. Individual event can be structured in many structures with increase rate of interactions and also may be game-based electronic learning. Simulation is another future of electronic learning. In healthcare electronic education, prices are lowered. We can sense that it will help in future about delivering learning content on exact demands. Mobile will be the most happening future of electronic education, where it can be merged with online real-time videos.

11.5.5 ELECTRONIC-DATA ENTRY

Numerous challenges are faced by healthcare sector. Unmaintainable charges, a rising non-insured patients and rinsing on ability due to better request are few issues. In these burdens, healthcare suppliers, clinics continually fight by offering excellence healthcare struggling with every-day's operative demands. This vibrant, dynamic realm sorts it imperious for health suppliers to establish better working methods. For example, method of allocating non-medicinal jobs of medical data entry services, etc., to a reliable retailer with low costs is risky. This method will help to focus on offering and taking care of health medical roles and the quality of system.

Experienced specialized retailers will ensure to get first class services, for example, Outsource2india. The jobs include: file data entry facilities, inline, image, and offline data entry facilities, data seizure and enrichment services, database expansion and relocation facilities, copy–paste facilities, manuscript organization systems, data removal facilities, volume data entry facilities, and directory facilities. These methods can be customized to get exact needs. Skilled data entry facilities are also available such as website data entry and database data entry facilities.

11.5.6 COMMUNITY OR GROUPS

People in various societies share their own knowledge of healthcare, its resources according to the needs. Native Broadcastings Corporation contributed extra area in thousands for text communications such as "Smile for You" and "Please Call Me" movement to deliver operation aimed at kids in South Africa for free care that helped 42 kids for surgical procedure.[24] Equally, society fellows care on activities of pre-emptive or management of a common disease. These social systems are formed between healthcare associations and persons used to exchange data. The threats in these methods are no interaction among limited groups and are directed toward over outflows and uncoordinated series that can direct to system inadequacy.

11.5.7 INDIVIDUALS

M-health for individuals provides improved interaction strategies to use, to diagnose, to save, and to retrieve personal health and medical information in central database. Advancement in cellular technology permits to transfer bulky information files such as medical imaging data for processing by local clinics to multi-specialty medical centers. This helps in perfect diagnosis and in providing better care.[18] For instance, establishing a zonal unwired network to maintain nonphysician contributors in 31 distributed eye clinics is done by Aravind Eye Systems (Madurai, India). First, individual patient is observed by the nurse and further proceeded for consultation with an ophthalmologist at the main medical center.[54] Patients confidentiality is an important factor to be maintained while transferring medical images.

Personal Digitals Assistant can also be used in this resolution.[48] Interaction and communication is most important aspect to deal with the patients for treatment of chronic diseases. M-health allows patients to put query and access reply, it sends information guidelines and keeps database of patients once patients leave the hospital. HIV/AIDS has advanced tools for use of M-health of patient requiring complex care. Evidences are collected and reported regarding the increase in improvement of such patients in their health and medication system by using M-health.[15,42] Self-activated drug observance reminders,[43] in-the-field consultations for providers,[14] and motivating healthy behaviors are other applications included in M-health.[19,20]

Health behavior and follow-up appointments are encouraged by using text message and phone reminders. In the United Kingdom, text message reminders are used for sexually transmitted infection heath centers to provide advantages such as lessening time to treat *chlamydia* and decreasing appointment no-show rates.[47,58] In Hangzhou, China, messaging cost less than telephone reminders and SMS-enhanced schedules presence is about 7%.[16] In Malaysia, a latest randomized measured test of patients with chronic complaints found that absence rates were about 40% lower in the SMS and mobile device community than in controls.[4,44]

11.6 MOBILE AND SOCIAL MEDIA USE: EMERGING TRENDS

The discipline of interactions has established varied feedbacks to the beginning of digitalized media few individuals and firms actively participating and involving with it, while others took time for both the fascination with digitalized media to depart and to observe achievement and discover modules by emerging their own approaches. The task of mobiles with text messaging has already been advanced in the field of journalism. Study also discovered that mobile text messaging leads toward major decline in occurrence of methamphetamine usage and insecure sex because outdated methamphetamine treatment leads to men-sex-with-men (MSM).[60]

There has been use of: blend of ProQuest, Google Scholar, and PubMed-published earlier data and grey literature, a blend of common web and targeted searches engines on relevant sites (e.g., organizations, associations, and academia) also broad search engines of transcripts journals, magazines, and newspapers. Common sorting and searching algorithms also were employed, which are grounded on emerging appropriate nomenclature that is associated with the major focus of the study, such as: digital and social media, Facebook, social networking, research health, and mobile health and media.

Further, database-precise explore factors and key words such as descriptors and subject headings are hired. Entire ties of 222 articles are fetched out among that 179 articles are reviews and the rest were grey literature. Range of publications dated from 2009 to 2013 of the document. The consequences of this review discovered numerous issues:

- First, transversely the early reports are understandable that digitalized media comprise center of fitness involvement activities, research studies, and campaigns. The literature is wealthy exemplified with the way mobile and social media strength can be utilized to broadcast consumers on health knowledge and eventually effect on its position.

- Secondly, a major cart off information is to find either they are healthy or unhealthy actions, and if digitalized network can be utilized effectively. Whatever behaviors can be stimulated depends on the users and with whom the messages are being interacted.[64] At last, in context of mobile and its technology whether communication was based on text-messages,[56] social media on location-based for instant FourSquare[28] or other smart-phone applications.[30] Mobile utilization is robust on heath research. The outcomes are capable for health interferences with outcomes showing better observance to terminate smoking, 15 declined in therapeutic medicine use and effective education on sexual health decisions,[51] whereas the use of digitalized media for interaction is mature through patterns of execution, there is need of efficient and significant assessment. Although there are numerous studies and majority of them is restricted labratory testing, there is a scope for establishing such experimentation in context of broader, difficult viewers, fledgling.
- Almost all viewers demoted to use the regular dimensions on basics of "likes," "click," and "tweet" data that are willingly accessible. Thus, need for additional research evidence in the field is mandatory.

Furthermore, as the world approaches toward usage of digitalized network in the field of health research work, the health society has been immediate to grant the digitalized media trend besides to construct various observations on digitalized network involvement also movements. However, it has been deliberated to expand useful approaches to utilize digitalized network for specific uses. The outcomes are the confirmation from this review: To greater extent, the earlier reports are centered on work involved on practical use of mobiles that will be carried out in future research. Furthermore, studies also explain optimistic outcomes in provisions of custody of contestants, together with continued learning process. Individual person can find 19 contestants that may have been lost in the study thus reducing erosion by 16%.[21] Based on this whole theme recognized, there emerges a key development for health and digital media. These are discussed in the following sections.

11.6.1 CONNECTING HEALTHCARE PROVIDER AND CONSUMERS

The capability to take notes online and watch latest developments or observe the signal[27] about its movement, product, or concern is mostly untouched in health sector, whereas novel observation is employed through[7] discussion is not captivating till now. However, this known information can be superimposed with other investigation or interactions, which promote to build up a profounder perceptive of occurrence. Therefore, innovating present practices can exploit the efficacy of the gathered information. This is achieved by dynamically pairing conventional schemes with this latest technology of online listening.

Social media significantly facilitates the ability to connect online with health researchers or health providers or policy analysts. This can be done from supports of "Meetups"[70] on meetup.com, wherever health-related programs may be able to toil with embattled Meetup community to grant appropriate health-related well-known news. Similarly, we can share useful health–fitness news on Tweet-Ups[6] or Twitter-Chats[49] via online virtual programs.

Tweet-Up can get the form of events such as: physical gathering, and conferences, which has a chance on behalf of a society of expected Tweeters to convene in a social context, frequently in initial period. Therefore, the virtual group networks can be effective for the growth of latest communities inside well-known administrations and clusters/communities.

11.6.2 EXPANDING TRADITIONAL AND NON-TRADITIONAL INFORMATION

Today and in the past, the healthcare information intended for the community has often been spread by call centers and depots of mass mailings.[59] These possibilities till now are important and applicable for numerous population, discovering methods to set the "public" support into "public health communication."[76] This can incorporate approved digitalize medium, which guides into conventional distribution performances. The "Best Buy model"[62] has altered the consumer check form by adding a Twitter-strategy through tweets queued-in. The priority of patients' treatments is vital as it can be understood as being practical instead of being imprudent to customer mails and phone calls.

Joint venture efforts have presented a key task in the endorsement of health issues, health-fitness care, and health conditions. On the other hand, the health room has inclined to continue associated with the same classes of allies—educational, investigate, and nonprofitable associations.[69] There are prospects to arrive at patrons with health-fitness information via conventional allies and the chance to use digitalized associates is a piece of these. Coordinating through pioneering digitalized associations includes About.com and WebMD.com that recommend prospects to set healthcare data in face of busy persons and line up topics of connected issues. The various sort of separate digitalized issues (e.g., post, forums, blogs, and social media stand) are also helpful medium to involve customers, other than falsify affiliations in various paths through digitalized corporations, which permit healthcare issues to distribute and allocate through confidence vote of superior and extremely directed viewers.

11.6.3 RELATIONSHIP AND MANAGEMENT AMONG SOCIAL MEDIA USERS

Social consumer relations management (CRM) is focused on thoughtful users in social media[31] to ensure that texting and its content are appropriate and attractive. Use of social media is a main form of interaction, which can quickly engage wider dispersion of larger electiveness and of extra conversation. Capability to realize goal or subgoal by user can obtain the maximum significant data that is hiding behind digitalized medium. In conventional health area, we are now deprived in dividing the user for each other's interaction. Generally, this has happened due to limited funds used for the costly procedure of recognizing, refining, and getting aimed users. However, with clearly cheap tools accessible via digitalized medium, there is a vast prospect for health care to acquire the true text to the true user.

11.6.4 E-HEALTH

An obvious operation in the health sector and along with teamwork of government of the United States, healthcare sector information is further unlocked and accessible to public and can comprehend to handle individual responsibility. Furthermore, an available witness is an outburst of mobile applications,[32] which influences the information to help patients.[10] We also have seen this happen too with crowd-sourced health information[25] and health-related information that is obtained from different groups, and online infirmity outbreak tracking systems. This is befitting common-place from corner to corner of the world with model like father in South Wales, the United Kingdom, who effectively make a diagnosis of his child over his Facebook social network[12] and the mother from the United States with an analogous story.[67] Subsequent step for health will be to slim it down to those of "exceptional applications" that really are worth to lives of people and estimating their latent to progress health fitness.

11.6.5 CONVERGENCE OF SOCIAL MEDIA AND MOBILE FOR HEALTHCARE

Among private/personal health data and the developing eminence of mobiles utilization of data and communication through mobile, there is an extensive prospect for the healthcare sector. Primary and for nonminority and minority crowd is same access at similar kinds of social media along with some regularity. It is well known that minorities use social online websites and adding data through their mobiles despite of having smartphones or having advanced tariff on landline, by means of mobile-socialized platforms, for example, GetGlue,[26] Instagram,[35] and FourSquare.[23] Data sharing and networking is always immediately obtainable[33] and the distance is not a limitation. This capability of retrieving data persists unexploited prospect for embedded health programs and is well known as "mobile social network."[1]

11.6.6 "BIG DATA" IN E-HEALTH

As stated earlier, available requirement for majority of scrupulous is a regular dimension related to digitalized media. Accessibility of information is understood as "Big data" (called as bulk of data) in digitalized medium.[68] However, one should develop the perceptive of what queries to ask about information and be conscious of how to respond.[65] As large databytes.com mentions, in latest era, it is about drawing out the price from the databases.[38]

11.6.7 EMERGING TRENDS IN DIGITAL COMMUNICATION

Almost it is obvious that plans in the digitalized medium inside associations cannot take long time to be handed over to trainees, subordinate, or hourly employees; and it frequently takes place on the source, where users are mostly youth. Moreover, sometimes there is disconnection among them, who have intentional control of healthcare institutes with excess juniors taking care of digital medium. Resulting toward escort to muffled messages (that are reserved in time scale) is also satisfied by needs of getting prior authorization from superiors. Requirement for private assurance of high-ranking control in digitalized interaction is supreme, and they are applicable to essential commercial values for the prosperous utilization of digitalized medium in healthcare sector. High-ranking influence will perceive job to become the identity with face or person's voice of their association on TV channels and radio channels. Therefore, they can boast toward to pick up fast skills of learning digitalized medium as a means of interaction with the public, long-suffering people, staff, and key stake-holders.

11.7 PREVALENCE OF SOCIAL MEDIA ENGAGEMENTS IN HEALTHCARE SECTORS

Social media is gradually becoming more ambitious by authoritarian pressures, and the first requirement is to reduce overheads and leftover alarm in the healthcare industry. Consumer opinion to progress trade has therefore become most important. Social media is establishing communications among end users and internet service providers that need simple access (and can also access social media through mobile devices) and impartial stands for distributing feedback. Slightly interested healthcare service providers across global, available on social networking websites are: YouTube, Facebook Twitter, and blogs.

11.7.1 IMPORTANCE OF SOCIAL MEDIA IN HEALTHCARE

The internet has key role to facilitate health information distribution, with social media as prime source. About 60 million Americans switched their medicinal needs to online timings compared to the previous year. Approximately 72% of sick persons seek out online data in before or after visiting to physician. About 890 medical centers in the United States are using social media to employ their patients indicating: 73% of US customers being healthy; 74% responding to be healthy. A predictable 500 million people globally were estimated to be using mobile healthcare applications in 2017. About 17,000 health-related applications existed till November 10, 2010 with more than half intended for patients instead of health professionals. The usage is about 15% among young adults (age 18–29) for medical apps in society, compared to 8% for adults (age 30–49). Other main healthcare movements in 2010 composed of: a lately accomplished customer assessment in the US emphasized about positive impact of social media in healthcare and healthcare social medium can be used for employment, distribution rates, launching latest facilities and products, and consciousness establishment. Let us come across at these models with a few present instances.

11.7.1.1 SPECIALIZED SKILLS RECRUITMENT

Ever since the trade needs of focused expertise, administrations have encouraged introducing advertisements in traditional method to advanced targeting social media channels, for example, alerted medical colleges sites, industry professional and skilled students blogs, and a variety of professional social websites, for instance, social media savvy and LinkedIn, associations wish to attain platform reachable of social network. As an instance, the HCA group (http://www.acareerathca.com/) enthusiastically uses twitter(@acareerathca) to publicize for vacate hospital locations.

11.7.1.2 ASSISTING CUSTOMER ASSESSMENTS

Health centers are upgrading blogs as means for consumer ranking and opinion. Few administrations also let users to blog on the blogs of manufacturer or connect the user's private blog to their cooperation site regarding their usage. To publish the Patient Satisfaction Ratings (PSR), Todd Linden (at Grinnell Regional Medical Center in Iowa) used blog belonging to him. However, the Hospital Consumer Assessment of Health-care Providers and Systems (HCAHPS), in cooperation with Centers for Medicare & Medicaid (CMS) and the Agency for Healthcare Research and Quality (AHRQ) are publishing online the information for users. Most of hospitals with higher assimilation of social medium in their advertising and communication strategy have prepared better compared to their peers in the HCAHPS. The fine model is Sarasota Memorial in Florida, which has 1230 followers on Facebook. It has gain of 6% over its opponents in the HCAHPA rates and is categorized as "Definitely would recommend."

11.7.1.3 AWARENESS CREATION

In a proposal toward penetrating into new marketplaces in developing countries, firms conventionally are dependent on eligible third person investigation to recognize tradition and its effect on trade capacity. At present, a huge segment of investigation is substituted by client deliberations on devoted sites. A good example is Johnson & Johnson's BabyCenter chain of sites that emphasis on health of baby and mother. Conversation themes consist of: rating/approving, tests through pregnancy, gynecologists, baby goods, opinion/comments on hospitals and services, customary birth/post-birth observance against current practices, and exercise practices before and after pregnancy.

The website was initiated as a pilot in the United Kingdom in 1997 and has now extended to 22 markets across the world. Another case is when numerous scientists at The Massachusetts General Hospital (MassGen) emerged with an apple-iPhone app for positioning neighboring critical ward wherever in the United States. Mass Gen, with an extremely established social medium strategy, shaped a YouTube video regarding the app and inclined it to bloggers to utilize the application. They too tweeted and posted on Facebook pages. It allows the application to list on social network websites such as: WorldCom's Geek Dad Blog and Boston.com.

11.7.1.4 MAJOR ISSUES IN HEALTHCARE SOCIAL NETWORK

Major apprehension via social network in trade has been health data. In deeply regulatory and usual determined production (e.g., health), it grows to obligate fine familiar/qualified bases of data. For example, a branch of HHS endorsement possibly will turn into distant extra precious than the hundred customer feedbacks on an item. Because of amplified privacy laws and regulation, healthcare-centric thoughts require superior control than other types of trade. This also widen to social media in provisions of describing obvious standards/strategies for quantity of topics to allocate.

11.8 DISCUSSION

The optimistic latent for M-health is enormous; however, present risks must be supervised and reduced. Quite a few extensive points tolerate prominence. First, careful evaluation is necessary. Proof for the worth of M-health remains limited, particularly for the developing world. Enduring estimation of definite schemes must direct on M-health growth. A combination of randomized measured tests,[43] normal experimentations, and further plans are desired. We consider that the fields will be in advantage really by the establishment of an M-health assessment registration, like registering at clinical trials at http://ClinicalTrials.gov: Service by the US National Institute of Health, which maintains a data bank of all clinical trials around the globe. Worldwide health organizations synchronizing bodies (e.g., World Health Organization, WHO) can contribute significantly by nurturing research studies using similar consequence procedures and publishing results.

Lastly, monetary results remain essential, such as: reduced costs, even absent enhanced health results, and might validate mobile-health methods. If mobile health is to be competitive with other health interpretations, then it must be quantifiable in provisos of expenses for each disability-adjusted life-year,[34] which can be acknowledged by health-intercession routine. A financial assessment structure for M-health would describe the intercessions, its expenses, and their proposed medical consequences and possible undesirable effects. Measurement of M-health must also comprise attentiveness of real-world matters, such as sustainability. Realistic dares intensely impact the ability of plans to stay alive in addition to mature.[9] Subjects involve organizations, technological ability, and backing. The organizations like WHO should build up assistance on best of M-health services and upkeep countries' accomplishment determinations.

Dynamic growth of M-health might possibly also have a non-health profit, for example, nurturing local economic expansion apart from healthcare. Even though scientific developments in hardware sometimes happen in developed countries, developments in software are not limited. Hence, M-health advancement and execution are doubtful to be restricted to outside professionals. M-health software apps will be planned and hired in intentional corporations with individual businesses, governments, health systems, communities, and professionals in developing countries. Such mobile platforms can be refined, expanded, and tailored in consultation with local professionals. To put into practice M-health, local technological capability and training is very essential. Thus, M-health must promote local microventure and local economy, generating advanced platforms and novel functionalities that can produce unending economic prospects. Finally, improved economic prospects in-turn can add to improved health. The "honest cycle" will go on, with M-health at the very core of it.

The value of communicating thoughts, on healthcare and assisting knowledge and argument among the public and interest groups is a lasting module of the performance of fitness structure around the globe. Same as TV and data lines have turned into vital a medium of interacting fitness alerts and notions, the initiation and thrilling development of digitalized media is the technique where fitness regarding data is interacted and congregated. Digitalized media have basically changed the communication style about health-related issues. Examples are: reflexive one-way move of information, fitness experts to the community, lawmaker whose borders with open active talks done, sometimes at quick speed in a whole day. Tendency to thoughts, data, and remark to distribute quickly on digitalize networks can generate chances to healthcare specialists and heads of communities to converse and teach innovative methods also attain public of concern.

Capability to collect information via digitalized network is emerging promptly. Further information is useful to keep record on spread of ailments and mobilize reactions to critical and rising troubles. The chance to create information obtainable on digitalize media in willingly reachable structures may be able to improve the capability of singles and societies to examine discussion and recognize health care upon innovative customs. Chance for public health organizations and health care to outline discussions and interact in a sensible manner with broad sections of the people exploiting digitalize medium may be persisted to mature. Toward getting highest gain of chances may need a new way of management. Leaders will have a genuine attendance in the field of digitalized medium with the self-assurance to communicate with digitalize network, Twitter, as subjects and discussions grow. Controlled expression of medium affairs also irregular publication rarely sufficient as the closeness, familiarity and accessibility of digitalize medium makes a fresh health in general a vibrant chance. Catalog of movements for digitalize medium exhibits versatile characteristics of digitalize medium with increased chances to perceive enhanced informed programmed scheduling, extension of health services, mature link structure, technology improvement, customization of health issues, further healthy and significant dimension and assessment, and finally, digitalize medium as the outstanding means for healthcare transformation.

11.9 CHALLENGES AND FUTURE DIRECTIONS OF SOCIAL MEDIA IN HEALTHCARE SECTOR

During mobile remote monitoring devices, mobile applications, sensors, and orientation resources, healthcare delivery can be enhanced by abundant opportunities. Discovery is a large part of this environment because it is tough to construct new hardware or software applications without extended atmosphere that supports and boost ingenuity. To give confidence to advance mobile health, many suggestions and numerous measures are intended to progress the acceptance of mobile medical devices and its applications. Policy-makers must promote the utilization of mobile devices for healthcare.

Tablets and smartphones have increased quickly in developing and developed countries, and this signifies a chief prospect to change the mode in which medical care is conveyed. Electronic systems for service delivery will offer advanced levels of quality and are inexpensive. Chronic diseases are an expensive and involve almost three quarters of medical costs. They comprise asthma, cardiovascular disease, cancer, and diabetes. We are supposed to support the use of mobile systems that can check patient symptoms and offer immediate recommendation to start the treatment, medication, and precautions because these have the possibility to manage minimum expenses, lower inaccuracies/errors, and progress patient's practices.

There are many mobile apps to help in identifying chronic disease, sensors, and remote plans that observe patient's physiology and electronic-libraries to fetch the newest information on health sources around the world. These supplies stand for a significant rise ahead in contribution superiority of healthcare. We ought to sort these tools that are much more broadly available and work to get rid of obstacles. The identical is accurate for clinical decision support for healthcare providers. With advent awareness about genetics, diseases, and pharmacological products, we are making the amenities of medicine far more complex. General practitioners are eager to gain information about newest discoveries in medicinal stream and refer such information to their patients. Software for facilitating health experts to comprehend how to deal during signs and what medication connections they must stay away from are gradually more being viewed in similar reference library than a medication tool and regulating the health issue. Health experts must access data on treating a variety of illness.

Furthermore, main obstacles to price suppression and quality of service delivery have been persistently trustworthy in several conditions on document-based medical systems. Medical doctors used to prescribe medicine on paper, lab tests reports are documented or pictured, medical accounts are stored in cabinets, insurance declarations are rewarded by repayment requests directed by the mail. In this digitalized world, individual cannot visualize an expensive means to execute a healthcare system.

On cross-examination of government instructions, the FDA has fixed its assistance on how mobile apps and structured mobile medical devices have demonstrated to elucidate some of the indistinctness and help additional modernization. Obligating apparent regulations to push popular activities is the top priority to shift further in mobile health. This means that lots of nations including the United States are being challenged to offer adequate healthcare. Complications comprise substantial space among doctors and patients, also untrained healthcare experts and the unexpected complication amid narrow medical systems and overheads of healthcare apparatus and infrastructure. Adding together, the recent epidemic of chronic illnesses depicts the required modern, well-organized, advanced technology-supported interference.

Mobile technology assists in information access to help: to lower expenses, rising efficiencies, simplifying inaccessible care, and by involving patients by contributors almost everywhere. Mobile health apps and their services are attracting vital devices in advancing healthcare assets all over the globe.

11.10 SUMMARY

M-health signifies a junction of healthcare and mobile skills. The M-health technology uses all qualified components, which are responsible for transmitting patient's medical information. Mobile technologies provide the capability to connect patients to their medical doctors, supervisor, and esteemed ones for timely medical supervising, which recommends enhanced patient commitment and improved health conditions. For using M-health, one should be equipped with mobiles, tablets, or equivalents to install and upgrade applications. Some knowledge about operating smart mobile can be helpful to use the health apps efficiently. Awareness about the mobiles, apps, and internet services in the developing nations has come up long way affecting positively in assisting medical health issues in quick time. In the meantime, some simple apps are part of the people's lifestyles, for example, diet, fitness, measurement of blood pressure, etc. M-health is a newly evolving subfield within E-health and M-services. Future dependency of M-health in developing nations, which is the focus in this chapter, will be full on the complete medical and mobile computing skills and health administrators and representatives and agencies or managers to establish conversant decisions on investments in health resources including technology.

KEYWORDS

- **developing countries**
- **digital communication**
- **healthcare sector**
- **mobile applications**
- **social media**

REFERENCES

1. Activ8soci al, Virginia, C2013. https://www.computerworld.com/article/3156932/mobile-wireless/darpa-wants-to-create-secure-data-sharing-tech.html (accessed Apr 9, 2018).
2. Adair, L. S.; Popkin, B. M. Are Child Eating Patterns Being Transformed Globally? *Obesity* **2005,** *13* (7), 1281–1299.
3. Adeyi, O.; Smith, O.; Robles, S. *Public Policy and the Challenge of Chronic Non-communicable Diseases*. International Bank for Reconstruction and Development, World Bank Publications, Washington (DC); E-book: http://hdl.handle.net/10986/6761; 2007; pp 218.

4. Agha, S. The Impact of a Mass Media Campaign on Personal Risk Perception, Perceived Self-efficacy and On Other Behavioural Predictors. *AIDS Care* **2003,** *15* (6), 749–762.
5. Arah, O. A.; Ogbu, U. C.; Okeke, C. E. Too Poor to Leave, Too Rich to Stay: Development and Global Health Correlates of Physician Migration to the United States, Canada, Australia, and the United Kingdom. *Am. J. Pub. Health* **2008,** *98* (1), 148–154.
6. Barnard, S. R. "Tweet or Be Sacked": Twitter and the New Elements of Journalistic Practice. *Journalism* **2016,** *17* (2), 190–207.
7. Bode, L. Facebooking It to the Polls: A Study in Online Social Networking and Political Behavior. *J. Inform. Technol. Pol.* **2012,** *9* (4), 352–369.
8. Boulos, M. N.; Brewer, A. C.; Karimkhani, C.; Buller, D. B.; Dellavalle, R. P. Mobile Medical and Health Apps: State of the Art, Concerns, Regulatory Control and Certification. *Online J. Public Health Inform.* **2014,** *5* (3), 229–232.
9. Boulos, M. N.; Wheeler, S.; Tavares, C.; Jones, R. How Smartphones Are Changing the Face of Mobile and Participatory Healthcare: An Overview, with Example from eCAALYX? *Biomed. Eng. Online* **2011,** *10*, 24–28, https://biomedical-engineering-online.biomedcentral.com/articles/10.1186/1475-925X-10-24.
10. Boutayeb, A. The Double Burden of Communicable and Non-communicable Diseases in Developing Countries. *Trans. Royal Soc. Trop. Med. Hyg.* **2006,** *100* (3), 191–199.
11. Buckingham, D. Media Education Goes Digital: An Introduction. *Learn. Med. Technol.* **2007,** *32* (2), 111–119.
12. Burke-Garcia, A.; Scally, G. Trending Now: Future Directions in Digital Media for the Public Health Sector. *J. Public Health.* **2014,** *36* (4), 527–534.
13. Capurro, D.; Cole, K.; Echavarría, M. I. The Use of Social Networking Sites for Public Health Practice and Research: Systematic Review. *J. Med. Inter. Res.* **2014,** *16* (3), e79.
14. Chang, L. W.; Kagaayi, J.; Nakigozi, G.; Galiwango, R. Telecommunications and Healthcare: An HIV/AIDS Warmline for Communication and Consultation in Rakai, Uganda. *J. Int. Ass. Phys. AIDS Care* **2008,** *7* (3), 130–132.
15. Chang, L. W.; Kagaayi, J.; Nakigozi, G. Effect of Peer Health Workers on AIDS Care in Rakai, Uganda: A Cluster-randomized Trial. *PloS One* **2010,** *5* (6), e10923.
16. Chen, Z. W.; Fang, L. Z.; Chen, L. Y.; Dai, H. L. Comparison of an SMS Text Messaging and Phone Reminder to Improve Attendance at a Health Promotion Center: Randomized Controlled Trial. *J. Zhejiang Univ. Sci.* B **2008,** *9* (1), 34–38.
17. Clarke, D. M.; Currie, K. C. Depression, Anxiety and Their Relationship with Chronic Diseases: Review of the Epidemiology, Risk and Treatment Evidence. *Med. J. Austr.* **2009,** *190* (7), S54.
18. Curioso, W. H.; Karras, B. T.; Campos, P. E.; Buendía, C. Design and Implementation of Cell-PREVEN: A Real-time Surveillance System for Adverse Events Using Cell Phones in Peru. In *AMIA Annual Symposium Proceedings;* American Medical Informatics Association (AMIA); 2005; pp 176–180.
19. Curioso, W. H.; Kurth, A. E. Access, Use and Perceptions Regarding Internet, Cell Phones and PDAs as a Means for Health Promotion for People Living with HIV in Peru. *BMC Med. Inform. Decis. Mak.* **2007,** *7* (1), 24–27.

20. Curioso, W. H.; Kurth, A. E.; Cabello, R.; Segura, P.; Berry, D. L. Usability Evaluation of Personal Digital Assistants (PDAs) to Support HIV Treatment Adherence and Safer Sex Behavior in Peru. In *AMIA Annual Symposium Proceedings;* PMID: 18999317; American Medical Informatics Association, 2008; pp 905–918.
21. Cuvelier, E.; Aufaure, M. A. Buzz and E-reputation Monitoring Tool for Twitter Based on Galois Lattices. *Int. Conf. Concep. Struct.* **2011,** *25*, 91–103.
22. Deb, L. Using Social Media to Reach Youth and Young Adults of Color, Personal Data, 2013. https://apha.confex.com/apha/140am/webprogram/Paper274959.html (accessed Apr 9, 2018).
23. Foursquare, New York, NY; C2013. https://foursquare.com/ (accessed Apr 9, 2018).
24. Free, C.; Phillips, G.; Watson, L.; Galli, L. The Effectiveness of Mobile-health Technologies to Improve Healthcare Service Delivery Processes: A Systematic Review and Meta-analysis. *PLoS Med.* **2013,** *10* (1), e1001363.
25. Gaba, D. M. The Future Vision of Simulation in Healthcare. *BMJ Q. Saf.* **2004,** *13* (Suppl 1), i2–i10.
26. Getglue.com (Internet). New York, NY: Adaptive Blue; C2013. https://telfie.com/?ad=getglue.com&utm_source=getglue.com&utm_campaign=getglue%2520404%2520redirect&utm_medium=redirect&utm_content=getglue%2520404%2520redirect (accessed Apr 9, 2018).
27. Greaves, F.; Ramirez-Cano, D.; Millett, C.; Darzi, A.; Donaldson, L. Machine Learning and Sentiment Analysis of Unstructured Free-text Information about Patient Experience Online. *Lancet* **2012,** *23* (380), S10.
28. Haithcox-Dennis, M. Foursquare: Health Education Specialist Checks-in-a Commentary. *Am. J. Health Edu.* **2011,** *42* (4), 194–196.
29. Hawkes, C. Nutrition Labels and Health Claims: The Global Regulatory Environment. *World Health Organization, Geneva,* **2004,** 1–88.
30. Hebden, L.; Cook, A.; van der Ploeg, H. P.; Allman-Farinelli, M. Development of Smartphone Applications for Nutrition and Physical Activity Behavior Change. *JMIR Res. Prot.* **2012,** *1* (2), e9.
31. Heller, B. C.; Parasnis, G. From Social Media to Social Customer Relationship Management. *Strat. Lead.* **2011,** *39* (5), 30–37.
32. Hopkins, P.; Hare, J.; Donaghey, J.; Abbott, W. Geo, Audio, Video, Photo: How Digital Convergence in Mobile Devices Facilitates Participatory Culture in Libraries. *Aust. Lib. J.* **2015,** *64* (1), 11–22.
33. Hostgator.com [Internet]; Texas; C2013. https://www.hostgator.com/blog/viewer-behavior-differences-on-mobile-devices-vs-desktop-computers/ (accessed Apr 9, 2018).
34. Iluyemi, A. Community-based Health Workers in Developing Countries and the Role of M-health. In *Telehealth in the Developing World;* Wootton, R., Patil, N., Scott, R. H. O., Eds.; Royal Society of Medicine Press/International Development Research Centre, London, UK, 2009; p 324.
35. Instagram.com [Internet], California. https://www.instagram.com/ (accessed Apr 9, 2018).
36. IOM.edu [Internet], Women, Infants, and Children (WIC). https://www.fns.usda.gov/wic/women-infants-and-children-wic (accessed Apr 9, 2018).

37. James, W. P. T. The Nutritional Crisis to Come. In *Critical Issues in Global Health;* Koop, C. E., Pearson, C. E., Schwarz, M. R., Eds; 1st ed; Jossey – Bass Publishers: San Francisco (CA), 2001; p xxvi.
38. Jamison, D. T.; Breman, J. G., Eds. *Disease Control Priorities in Developing Countries*, 2nd ed.; Oxford University Press and World Bank: New York – NY, 2006; pp 3–34.
39. Kaplan, W. A. Can the Ubiquitous Power of Mobile Phones be Used to Improve Health Outcomes in Developing Countries? *Global. Health* **2006,** *2* (1), 9–11.
40. Krishna, S.; Boren, S. A.; Balas, E. A. Healthcare via Cell Phones: Systematic Review. *Telemed. J. E-Health.* **2009,** *15* (3), 231–240.
41. Lansing, A. E.; Washburn, J. J. Cognitive and Academic Functioning of Juvenile Detainees: Implications for Correctional Populations and Public Health. *J. Correc. Healthcare* **2014,** *20* (1), 18–30.
42. Lester, R. T.; Gelmon, L.; Plummer, F. A. Cell phones: Tightening the Communication Gap in Resource-limited Antiretroviral Programmes? *Aids* **2006,** *20* (17), 2242–2244.
43. Lester, R. T.; Mills, E. J.; Kariri, A. The HAART Cell Phone Adherence Trial (WelTel Kenya1): A Randomized Controlled Trial Protocol. *Trials* **2009,** *10* (1), 87.
44. Liew, S. M.; Tong, S. F.; Lee, V. K. Text Messaging Reminders to Reduce Non-attendance in Chronic Disease Follow-up: A Clinical Trial. *Brit. J. Gen. Prac.* **2009,** *59* (569), 916–920.
45. Lopez, A. D.; Mathers, C. D. Measuring the Global Burden of Disease and Epidemiological Transitions. *Ann. Trop. Med. Parasitol.* **2006,** *100* (5–6), 481–499.
46. Mathers, C. D.; Loncar, D. Projections of Global Mortality and Burden of Disease from 2002 to 2030. *PLoS Med.* **2006,** *3* (11), e442.
47. Menon-Johansson, A. S.; McNaught, F.; Mandalia, S.; Sullivan, A. K. Texting Decreases the Time to Treatment for Genital Chlamydia Trachomatis Infection. *Sex. Trans. Infect.* **2006,** *82* (1), 49–51.
48. Mitchell, M.; Getchell, M.; Nkaka, M.; Msellemu, D.; Van Esch, J.; Hedt-Gauthier, B. Perceived Improvement in Integrated Management of Childhood Illness Implementation Through Use of Mobile Technology: Qualitative Evidence From a Pilot Study in Tanzania. *J. Health Comm.* **2012,** *17* (Suppl 1), 118–127.
49. Moorhead, L. L.; Holzmeyer, C. In an Age of Open Access to Research Policies: Physician and Public Health NGO Staff Research Use and Policy Awareness. *PloS One* **2015,** *10* (7), e0129708.
50. Mullan, F. The Metrics of the Physician Brain Drain. *New Engl. J. Med.* **2005,** *353* (17), 1810–1818.
51. Mychasiuk, R.; Benzies, K. Facebook: Effective Tool for Participant Retention in Longitudinal Research. *Child Care Health Develop.* **2012,** *38* (5), 753–756.
52. Nisbet, E. C.; Stoycheff, E.; Pearce, K. E. Internet Use and Democratic Demands: A Multinational, Multilevel Model of Internet Use and Citizen Attitudes about Democracy. *J. Comm.* **2012,** *62* (2), 249–265.
53. Non-communicable Diseases and Their Risk Factors [Internet]. http://www.who.int/ncds/en/ (accessed Jun 30, 2018).
54. Olayiwola, J. N.; Sobieraj, D. M.; Kulowski, K.; Hilaire, D. S.; Huang, J. J. Improving Diabetic Retinopathy Screening through a Statewide Telemedicine Program at a Large Federally Qualified Health Center. *J. Healthcare Poor Unders.* **2011,** *22* (3), 804–816.

55. Ozdalga, E.; Ozdalga, A.; Ahuja, N. The Smartphone in Medicine: A Review of Current and Potential Use Among Physicians and Students. *J. Med. Inter. Res.* **2012,** *14* (5), e128.
56. Perry, R. C.; Kayekjian, K. C. Perspectives on the Use of a Text Messaging Service for Preventive Sexual Health Promotion. *J. Adolesc. Health* **2012,** *51* (3), 220–225.
57. Popkin, B. M. The Nutrition Transition and Obesity in the Developing World. *J. Nutr.* **2001**, *131* (3), 871S–873S.
58. Price, H.; Waters, A. M.; Mighty, D.; Nixon, J.; Burton, N.; Picket, J.; Sullivan, A. K. Texting Appointment Reminders Reduces 'DID Not Attend' Rates, Is Popular with Patients and Is Cost-effective. *Int. J. STD & AIDS* **2009,** *20* (2), 142–143.
59. Raja, J. D.; Bhasin, S. K. Health Issues among Call Center Employees, An Emerging Occupational Group in India. *Ind. J. Comm. Med.* **2014,** *39* (3), 175–180.
60. Reback, C. J.; Grant, D. L. Text Messaging Reduces HIV Risk Behaviors Among Methamphetamine-using Men Who Have Sex with Men. *AIDS Behav.* **2012,** *16* (7), 1993–2002.
61. Rogers, C. R. Social Media, Libraries, and Web 2.0: How American Libraries Are Using New Tools for Public Relations and To Attract New Users. *South Carolina State Documents Depository,* Clemson – SC; 2009, Mar 22; p 30.
62. Ryan, D. *Understanding Digital Marketing: Marketing Strategies for Engaging the Digital Generation*; Kogan Page Publishers: New Delhi, 2016; pp 125–235.
63. Scheffler, R. M.; Mahoney, C. B.; Fulton, B. D.; Dal-Poz, M. R.; Preker, A. S. Estimates of Healthcare Professional Shortages in Sub-Saharan Africa in 2015. *Health Aff.* **2009,** *28* (5), w849–w862.
64. Seidenberg, A. B.; Rodgers, E. J. Youth Access, Creation, and Content of Smokeless Tobacco ("Dip") Videos in Social Media. *J. Adolesc. Health* **2012,** *50* (4), 334–348.
65. SGI.com [Internet], California: SGI. C2009. http://www.youtube.com/watch?v=ayqIl Qr48P8 (accessed Apr 9, 2018).
66. Sheiham, A. Closing the Gap in a Generation: Health Equity Through Action on the Social Determinants of Health: Report of the WHO Commission on Social Determinants of Health (CSDH) 2008. *Comm. Dent. Health* **2009,** *26* (1), 2–13.
67. Slate.com [Internet], Washington DC, C2011. https://www.facebook.com/adhdcar-diff/ (accessed Apr 9, 2018).
68. Social Fresh [Internet], New York, C2018. https://www.socialmediaexaminer.com/ social-media-predictions-2018/ (accessed Apr 9, 2018).
69. Stuart, B. For-Profit Enterprise in Healthcare. *JAMA* **1987,** *257* (13), 1813–1815.
70. Tang, D.; Wei, F.; Qin, B. Building Large-scale Twitter-specific Sentiment Lexicon: A Representation Learning Approach. Proceedings of COLING 2014, the 25th International Conference on Computational Linguistics: Technical Papers, 2014; pp 172–182.
71. Tatsumi, Y.; Ohkubo, T. Hypertension with Diabetes Mellitus: Significance from an Epidemiological Perspective for Japanese. *Hypertens. Res.* **2017,** *40* (9), 795–799.
72. Taylor, A.; Chaloupka, F. J.; Guindon, E.; Corbett, M. The Impact of Trade Liberalization on Tobacco Consumption. In *Tobacco Control in Developing Countries;* Jha, P., Chaloupka, F., Eds.; Oxford University Press: New York, 2000; pp 343–364.
73. The Top 20 Valuable Facebook Statistics. https://zephoria.com/top-15-valuable-facebook-statistics/(accessed Apr 9, 2018).

74. Univision.com [Internet], New York: *Hispanic Patient Journey Study Identifies Gaps – from Awareness and Information to Diagnosis and Adherence.* https://www.businesswire.com/news/home/20120215005302/en/Univision%E2%80%99s-Hispanic-Patient-Journey-Study-Identifies-Gaps (accessed Apr 4, 2018).
75. Visually.com [Internet], New York, C2017. https://visual.ly/community/infographic/social-media/global-social-media-statistics (accessed Apr 9, 2018).
76. Viswanath, K. Public Communications and Its Role in Reducing and Eliminating Health Disparities. Examining the Health Disparities Research Plan of the National Institutes of Health: Unfinished Business. Washington, DC: Institute of Medicine, 2006; pp 215–253.
77. Wilson, M. G.; Ellison, G. M.; Cable, N. T. Basic Science Behind the Cardiovascular Benefits of Exercise. *Brit. J. Sports Med.* **2016,** *50* (2), 93–99.
78. YouTube.com [Internet], YouTube for Press. https://www.youtube.com/intl/en-GB/yt/about/press/ (accessed Apr 9, 2018).
79. Zoha, R.; Kumaran, S.; Hasmah, Z. SM Analytics: Impact of SM Engagement Metrics on Online Purchase Intention. *J. Eng. Appl. Sci.* **2017,** *12* (2), 283–289.

PART III

Recent Strategies for Mobile Phones and Social Media in the Economic Sector

CHAPTER 12

Potential of Streaming Transmission to Mobile Phone as Reinforcement of Community Radios

CELESTINO V. JOANGUETE

ABSTRACT

This chapter analyzes the relationship of convergence and integration of radio in the mobile phone and its transmission over the Internet. Authors emphasize how the *community radio* can be saved from the technological discontinuity and from the fragilization provoked by the systems of political regulation. The focus of this study is community radio.

12.1 INTRODUCTION

Technological changes are profoundly altering the history of radio, through an evolutionary process called radiomorphosis, which reconfigures the form of transmission and reception of contents. Consequently, two ruptures mark the history of radio. From the technological point of view: the first is the end of emissions through Hertzian waves and the second is the phenomenon of webradio that happens now with the transmission from the Internet to multiplatform digital content reception.

Mozambique has 110 community radios (CRs) with a mission of bringing information to rural communities. This represents 80% of the national audience. CRs greatly exceed the audience of other commercial radios. Religious nature is mostly concentrated in the capital of the country Maputo.

When talking about CRs in Mozambique, the radios that face operational problems (such as signal transmission, equipment quality, human resources, and circulation resources) are associated with the citizen's imagination. In addition to the problems that preceded the lack of a specific legal framework, there are other obstacles to its functioning since CRs are managed by community personnel without any technical training, which makes them the most fragile media in the national media scene.

This research is based on the following question: How can a CR transmit via streaming to the mobile phone as a way to overcome the crises of technical and financial sustainability? The questionnaire covers other small issues related to the end of airwaves, new audiences, geographic coverage, and cost of transmission in the webradio environment for the mobile phone.

In recent years, there has been a growth of mobile telecommunications sector in Mozambique, due to investment by operating companies. More than 19,223,516 users joined mobile telephone services, representing a penetration rate of 26.02%, as well as the geographic coverage of the entire national territory.[7] This scenario fuels the hope that radios can grow thanks to the strong power of distribution of the radio signal by companies of mobile phone.

In the light of vast literature on mobile telephone in Africa, the mobile media services at sub-Saharan African newspapers (a guide to implementing mobile news and mobile business) is the base of this study, which synthesizes detailed knowledge on the articulation between the radio and mobile phone in the technical process of transmission and dissemination of information.

The revision of this and other complementary literature, and taking into account the transformational phenomena that result from the mobile communication in alliance with the webradio, it is possible to understand the potential of mobile phone operators to provide new impetus for transmission and enlargement of the audience. The reason why we understand that it is necessary to study these new technological platforms of communication, it is hardly explored by the media, in parallel with the beliefs and expectations of citizen engagement within the Mozambican society.

To respond to the research objectives, authors reviewed the literature, reflecting on the multiple functionalities of the mobile phone, discussing the World Bank reports and interpretation of specific literature as well as analyzing the statistical data produced by the National Institute of Statistics of Mozambique on the penetration rate and ownership of mobile phone in Mozambique.[17] Authors did the counterpoint with CRs, emphasizing the combination between the mobile phone and the CR, in a logic of realizing that the streaming transmission to the mobile phone offers potential to reinforce the CRs.

This chapter explores the potential of streaming through the mobile phone as a way to reinforce the technical sustainability of CRs.

Once the technical weaknesses of transmission and functioning of CR were identified in this study, authors analytically reflected on the concept of convergence pointed out by Jenkins.[8] From this point of view, the study explored how to strengthen the continuity of radio in the digital era, in which it seeks a greater understanding of the potential of mobile phones and accessibility to the rural community. It is intended to understand the contours of the relationship between radio and mobile phone, transmission processes, and citizen participation, paying special attention to the changes that can transform it into a citizen engaged in society.

The objective of this research was to investigate the sustainable way of transmitting CRs via streaming to mobile phone. Therefore, the authors decided to adopt the qualitative and exploratory method of research. The research is exploratory because it is carried out in areas for which there is little knowledge accumulated and systematized.

This chapter is mainly of exploratory nature through theoretical discussions and field research, which results from an exercise that served to deepen the understanding of the subject and to know the specific reinforcements of mobile communication in CR, namely, on the contributions that it represents for the consolidation of citizenship. In this sense, authors have discussed two fundamental scenarios: (1) the idea that mobile phones are bringing communicational changes in Mozambican rural communities due to the investments on telecommunication infrastructures of mobile telephone companies on the Mozambican rural areas and (2) some streaming platforms for the mobile phone.

12.2 MULTIPLE PHONE FUNCTIONALITY

The United Nations sponsored the ECLAC studies entitled *La Nueva Revolución Digital*, in June 2015. According to this study, there were nearly 3.6 billion people subscribing to mobile telephonic services in the world and 2.923 billion persons using the Internet.

The massification of social technologies and telecommunications infrastructures has increased the global flow of communication.[1,2] As a result, it reduced costs of transactions, production, and distribution of digital goods and services.

In the African continent, mobile phone access rose from 8 to 100 million users, an increase of 107% a year from 2000 through 2005. The country with the highest growth rate was the Republic of South Africa with 34 million mobile phone users in 2005, compared to 9 million in 2001. For example in Botswana, every three persons have a mobile phone, while the Democratic Republic of Congo, Rwanda, and Malawi are investing in mobile communications infrastructure facilities, particularly mobile networks. These data confirm that African countries are experiencing accelerated growth in the use of mobile phones.

In the context of Mozambique, the adhesion to the services of the mobile telephone has grown enough in recent years. There were 12,808,480 subscribers in 2012 compared to 19,223,516 in 2013.[7] This growth is associated with the entry of new operators such as Vodacom Mozambique and Movitel.

In both Africa and Mozambique, one of the factors of the massification of the telephone by the population is due to the expansion of telecommunications infrastructure in the rural areas and the consequent competition among companies that supply services, as well as the affordable cost of the mobile devices from China.

The World Bank, in the context of the implementation of Climate Change projects and strengthening municipal capacity to strengthen urban infrastructures, applied the mobile phone to monitor climate change effects. The report produced by World Bank in 2015 on climate change in Mozambique indicates that resilience to climatic conditions in Africa is being strengthened by the use of mobile phones, particularly to warn governments of local problems of sanitation and drainage as well as how to map communities.[17]

Other examples of mobile succession in Africa are various cases of its application in the social field. In particular, there is a growth in payments for services and bank transfers, literacy, governance monitoring, and media reinforcement. This growth means that mobile services are rapidly reaching rural communities, previously marginalized by formal and bureaucratic systems.

In the political field, Benjamim Barber (1987) in his book entitled "Strong democracy: participatory politics for a new age" defends the principle that the flow of information through the Internet, including the use of the mobile phone, as a communication and interaction tool for public participation based on digital participation, has the same requirements for the operation of Benjamin spaces (https://www.ucpress.edu/book/9780520242333/strong-democracy).

From multiple functionalities of the mobile phone, two inferences can be drawn from this reflection: the device not only has the function of interpersonal communication, financial transactions, payments of services, monitoring and governance but it also has other potentialities that authors explored in depth: the possibility of articulation between the mobile phone and the CRs.

12.2.1 COMMUNITY RADIOS

CRs are radios that belong to the community in which the members are owners and decide on the content to be diffused. They promote citizen participation and defend their interests and their programs are debated and considered all of a view, in a context in which all are equally respected and considered.[2] CRs in Mozambique, in addition to conflicts with government officials that often dictate the closure and threats of their journalists, present numerous technical problems of transmission, namely:

- constant failures of the broadcasters;
- cost of equipment maintenance;
- obsolescence of equipment, inadequate technical assistance;
- poor coverage;
- poor financial resources; and
- poorly trained human resources.

In addition to the aforementioned problems, CRs have a reduced radius of coverage, which makes it difficult to reach a wider audience. Despite these technical and human limitations, they play a key role at the national level and in rural communities in particular.

In the Mozambican media scene, CRs have greater national coverage of about 80% of the audience. They are defined as nonprofit mass media, owned by communities where they are inserted and that autonomously manage and determine the contents and formats of the respective programs.

CRs in Mozambique are means of community development and social transformation through direct participation of their communities, the creation, and transmission of the message of social development in general, particularly those communities that are in need and far from urban centers.

Among 28 million Mozambican inhabitants,[7] at least 18 million of the Mozambican rural population listen to CR, with provinces of Zambezia and Nampula contributing with more listeners. In numerical terms, the province of Nampula has topped the list with a total of 18 CRs in operation.[12] The data presented here are reliable indicators that CR dominates the audience of the Mozambican rural population, particularly in the densely populated provinces. However, the data allow us to infer that in the national media scene, CR is the media of a major national audience.

With the transition from the analog broadcasting system to the digital broadcast, CRs will face challenges in the near future. Although the digital system is a real revolution in broadcasting, especially regarding the quality of the signal reaching the listener, yet they need to overcome their previous analog problems.

In several countries where the subject of digitalization of broadcasting is a matter of public discussion among experts, the government, radio managers, professional associations, scientists, technicians, and academics approach the subject by creating or adapting legislation to challenges of the digital age. At the end of the discussion on legislation, specialized commissions study and advise decision makers and the market on the digitization options, which enable governments to make timely decisions in a timely manner.

In contrast to other countries, there is little concern in Mozambique about the 110 CRs scattered throughout the country, all of which operate under the Associations Law 8/91 of June 18, and the Press Law 18/91 of August 10.

The law of associations ensures the freedom to create the association as a basic guarantee of the personal fulfillment of individuals in society and is established in Section I of Article 76 of the Constitution of the Republic as one of the fundamental freedom of citizens.

Another legal framework that ensures the functioning of the CR is the press law, which is the main legal instrument that regulates the exercise of the press in Mozambique, which gives freedom to Mozambican citizens to create media.

As Hill's[6] studies affirm in the context of Africa in general, the lack of a specific legal framework for CRs in Mozambique constitutes legal fragility for its functioning. The government recognizes its importance in community development and the democratization of society through the dissemination of information.

As can be seen, the importance of CRs in the development of the communities, where they are inserted, has encountered technical transmission difficulties, and this has made it extremely important to reflect on and debate about the possibility of finding a way to reinforce in the transmission to the stop of the discussion on specific legislation able to regulate the participation of CRs in development processes.

12.2.2 REINFORCEMENT OF RADIO TRANSMISSION VIA MOBILE PHONE

The 2008 publication by Lee Rainie, entitled "The Future of the Pew Internet Project," prognosis the dominion of the mobile phone as the main means of communication and of Internet access.[10] In fact, nowadays, the mobile phone is one of the appropriate devices for the majority of the world's population, particularly in Africa. In addition, it is one of the devices that materialize the concept of technological convergence with its multiple functionalities, which has attracted the presence of the media (television, radio, newspapers, and Internet social networks), and has consequently defined new forms of reception of the signal from the radio.

It is clear that contemporary society is witnessing the greatest phenomenon of convergence that is transforming relations between industry, market, audience, and consumption of news content products. An example of this is the convergence of the media in the mobile phone that represents current technological development.[8,9]

The convergence has allowed the radio to open new ways of transmitting their programs and to define the new concept of radio with absolutely different characteristics from traditional radio.

The use of mobile phones, tablets, and other mobile devices creates a closer proximity to the listener in the communicative process and, at the same time, allows them to be transmitted outside their traditional business and broadcast environment, which introduces transformations in the product consumption model radio.

The reception of the signal of the webradio depends, on the one hand, on a simple mobile phone with the appropriate application of reception of the audio. On the other hand, it also depends on smartphones capable of capturing or listening to more than 10,000 broadband radio stations. Therefore, the challenge is to build a simple interface that attracts cyberlisteners. The device must contain a radio-player application that is easy to use and can be modified according to the need and taste of the user.[9]

Till today, radio-scanning has occurred through the changeover of the terrestrial transmission delivery system from analog to digital. However, all new scanning initiatives face the same problem of lack of reasonably priced receivers.[9]

Cellular radio, without frontiers, is becoming a powerful medium in the rural context, due to low price and massification of devices to low-income population and limited financial resources. Therefore, there is a need to study the appropriate technologies that will allow it to reinforce the geographical reach of its transmission in simple and inexpensive ways.

For the purpose of searching for radiosignal transmission technologies and describing how they are operationalized, some game questions are fundamental, such as what are virtual radio creation platforms? What are modern forms of radio transmission to extend the range of coverage to the global scale?

However, broadcasting in streaming is a simple and inexpensive way of radio broadcasting.

12.2.3 STREAMING

Streaming is one of the technologies that can converge various media to an Internet environment. Streaming broadcasting emerged in 1997 and today it is one of the most innovative technologies in use on the Internet. It enables fast and continuous streaming of audio and/or video streaming over the Internet. This technology also allows transmitting the files at very high speed in an uninterrupted way because streaming intelligently calculates the quality of the link for upload and download starting the transmission only when the transfer rate is adequate.

It is up to the streaming server and the software installed on it to guarantee the quality of the transmissions, whether they are recorded or live. For each transmission, the server administrator has to create a link that is made available on a webpage and determine the number of users that will be able to access the transmission simultaneously.[11]

Streaming is used in audio and/or broadcasts of radio and TV stations, where all programming is displayed simultaneously via the web or via smartphones, tablets, mobile phones, and other mobile devices. The advantage of this system is that it reduces the transmission cost of radios and televisions with limited technological and financial sustainability capabilities. Streaming technology allows live streaming or streaming. Live broadcasts can be audio podcasts for AM and FM radios or video for broadcasters (web-TV or IPTV).

The *Radio Garden Platforms* were developed by the German companies Studio Puckey and Studio Moniker in partnership with the Dutch Institute for sound and vision. It allows, in the same space, to tune all radios that are transmitted by the Internet in streaming.[15]

The sites that concentrate radios are represented by green dots that when clicked allow accessing the live transmission of one of the stations of that locality. The rest is in the lower right corner to be accessed.

In addition, a link to the station's page is still available in the upper right corner of the map.

In the Radio Garden tuner streaming, there is a weak presence of Mozambican analog radios.[15] Only five radios are present, namely, LM Radio 87.8; FM Mozambique Radio 97.9; Radio Maria Mozambique; Maxaquene Community Radio, and Provincial Issuer of Gaza. Two are of public nature, one religious and one CR.

To enter the radio stations in the Radio Garden system, one must send a request by e-mail to jonathan@puckey.studio, with the following information: (1) name of the radio, (2) radio website, (3) radio city, and (4) URL of the streaming in MP3.

The presence of Maxaquene CR is already indicating that it is possible to transmit a CR with weak technical and financial resources to a global scale;[5] however, it is necessary to mobilize other CRs to enter the virtual world of transmission.

In another platform called Surfmusic, authors identified the presence of nine radios to be streamed. The nature of the radios that are transmitting in this platform is all commercial based in the city of Maputo.

To perform streaming, the user must have some multimedia equipment connected to the Internet, such as computers, smartphones, mobile phones, MP4/MP5 players, digital TV, or any other medium that supports the streaming link connection. One will also need the transmission execution software that is called a player.

Therefore, the massification of the mobile phone is essential for FM radio stations to be present on mobile devices. Fundamentally, it is necessary to encourage the population to opt for a mobile phone with radio application, to facilitate the listener to listen to their favorite radio on the cell phone, without the use of data from the Internet.

The potential of the mobile phone in transmitting the radio signal demonstrates the efficiency and reach that classic radio could not. The streaming transmission has established a fully differentiated broadcast structure that is richer in audience reach due to the Internet. However, for effective transmission of the radio in digital platforms it is necessary to requalify them for an autonomous transmission transmitter originating from the Internet, webradio. A process that requires all transformation to the virtual through the process called "radiomorphosis."

It should be noted that on the Internet, there are three categories of radio.[13] Each of the radios has its characteristics.

- *Offline radio station* is available on the Internet, but it does not run any sound program on an ongoing basis. They are only webpages that have institutional information and their programming. In this category, they use the website as a channel of interaction with listeners, where the listener participates in the programming through chats and e-mails.
- *Online radio* consists of online radio that is different from the previous ones. They provide the audio signal on the Internet and have a wider audience width and signal coverage.
- *Webradio* is exclusively for the Internet environment without any physical presence. This category of radio will delve into the webradio platform to better understand its operation.

12.2.4 PLATFORMS OF WEBRADIOS

Nowadays, the mass media, the broadcast has become a more simplified act of transmitting something, using any type of media, either through the Internet, radiowaves, satellite, cables, optical fibers or lines phone calls, etc. Transmission via the Internet allows sharing almost everything with others (text, video, and sound) due to various tools of diffusion such as webradio.

The classical transmission infrastructures present many problems of FM reception, namely, noise and loss of audio signal due to the presence of obstacles or the geographic location of poor signal coverage, as well as of all the political regulation. Webradio in the World Wide Web of computers, webradio like a transmitter that native of the Internet, amplifying the radio products and services.

Prata and Martins[14] indicated that the advent of the Internet has reconfigured radio broadcasting. The radio now has a webpage, with information about the company, programming, music, commercial, podcast, etc. Gradually, they also offer to stream, that is, radio content is being produced, transmitted, and accessed simultaneously on the radio, mobile phone, and computer.

With the development of the Internet, the web environment has become conducive to broadcast. Radio Klif, in Texas, the United States, was the first broadcaster to broadcast continuously and live through the Internet, starting in September of 1995. Between April 1996 and April 2000, the number of broadcasters broadcast over the Internet jumped from 56 to 3763 in the United States.

Webradio, in addition to the technology itself, opens the door to greater interactivity, offering new possibilities for broadcasters to interact with their listeners through SMS messages, program downloads, chat conversations, discussion forums, etc.

Although there are several methods of making and transmitting radio programs, yet it has evolved greatly in terms of technology and audience relations. Interactivity is one of the characteristics that brought radio closer to the listener, making it the medium of communication of the future.

Interactivity is understood as the interface between webradio and its cyber-listeners. According to Esteban and Alaejo,[4] it is a relationship mediated by new signal receivers, such as smartphones, tablets, YouTube, iTunes, Spotify, personal computer, and other mobile devices. Macedo[11] explains that to mount a WiFi, the first option is to "install, configure a WiFi system already developed, and host it on a server with streaming and auto DJ (used for even when one is offline)." The second, the author talks about transforming the personal computer or mobile phone into a server. This requires the installation and configuration of streaming and all other applications. However, for this, the computer must be connected to the Internet permanently.

The second step, the installation of the webradio requires the contracting of a provider/server, preferably of a plan with the stream with the availability of players so that webradio is heard. It is around the path opened by webradio in the digital ecosystem and which confirm its strength as media in the process of mediamorphosis number went as follows:

12.2.4.1 AIRTIME

Airtime is a platform for the creation of webradio, it was developed by SourceFabric previously under the name of Campster. However, in 2011, the company decided to open the program allowing the community to improve it. The Airtime platform has many functionalities, such as organized interface, organization of music, creation of playlists, music research, schedule.

The Airtime studio allows one to create a library with audio files stored in blocks that supports file formats, such as MP3, Ogg, WAV, FLAC, and AAC, that can be dragged into the playlist.

On the radio's webpage, anyone can quickly navigate the schedule. In addition, Airtime allows the editing of the programs and schedule them for automatic playback in days, week, month, and daily repetition, as well as supporting a live stream.

12.2.4.2 SAM BROADCASTER

This platform has the functionality to run an Internet radio station from a single computer. It can be used to run live programs with a DJ by selecting tracks to be played and mixing sound as well as fully automated, where programming is controlled by several sets of rules outlined by the operator.

SAM Broadcaster[16] connects to the most common broadcast servers, such as SHOUTcast and SHOUTcast2, as well as some web servers, such as Live365 or the company's paid broadcast, CheapestStream and SpacialNet hosting services. The program has a virtual studio with all the transmission features.

SAM broadcaster in addition to similarities with the Airtime platform offers the advantage of software from the clouds, even when the radio operator is absent.

12.2.4.3 BROADCAST MYSELF

Transmitting radio through a mobile phone has always been a problem, due to the lack of applications available with this feature. Broadcast Myself is a free Android application that supports radio stations. The system allows the live transmission of a radio from the microphone of the mobile phone.

The Broadcast Myself mobile radio application, relative to the other applications, has innovative and different transmission characteristics from the previous two platforms.[3] In addition to transmitting live, record, and save in MP3 format, it has a mixing console, controls the volume, and runs on the mobile phone functions similar to those of a radio studio.

The digital era requires radios to become present in the digital ecosystem so that they reach listeners. The implantation of a webradio, mobile application, social networks, live broadcast, SMS, and interactivity in the webpage attract the listeners and expand the time of their stay listening to the radio.

Airtime, SAM broadcaster, and Broadcast Myself systems and other streaming radio broadcast applications require a stable mobile phone connection to establish and maintain the open stream without loss of network.

As it can be seen in this study, the examples of webradio applications presented here can both migrate from a Modular Frequency FM radio to WiFi and use the mobile phone as a receiver as a radio studio can be created on it.

12.3 DISCUSSIONS

Based on the results of the General Reconstitution of Population in 2017 presented by the National Institute of Statistics, the study revealed interesting data on the penetration rate of the mobile device in Mozambique, as well as the increase in the number of service subscribers. These data allow us to infer that the mobile phone is one of the most appropriate social technologies for all Mozambican social classes, whether high and urbanized or the lowest and most ruralized.

From the social[17] and political (Barber: Section 12.2) point of view, the mobile phone is becoming one of the communication tools of great impact in the communicative, financial, and democratization process of information and, consequently, one of the key participation in the revolutionary process of production and distribution of information in the digital age.

In contrast, traditional media such as CRs, which have 80% of the audience of the 28 million inhabitants of Mozambique,[7] face numerous technical transmission problems, namely, (1) poor coverage, (2) constant failures of broadcasters, (3) poor financial resources, (4) cost of equipment maintenance, (5) obsolescence of equipment, inadequate technical assistance, and (6) poorly trained human resources.

Besides problems of a technical nature, CRs in Mozambique face problems of legal nature, which constitute fragility for their functioning, since the government is less concerned about this community environment, although it recognizes its importance in community development and in the democratization of society. Therefore, how to save CR from bankruptcy in the media landscape?

Both Jenkins[8] and Kemppainen[9] mention some trends of technological evolution that can rescue the radio from the technological discontinuity:

- First, the concept of technological convergence referred to by Jenkins[8] contains the indicator of radio integration in multiplatforms of production, transmission, and distribution of contents. These platforms allow one to integrate or create a radio with global broadcast capability, due to the streaming system.
- Second, Kemppainen[9] raises the issue of radio scanning so that it does not go bankrupt. However, these radio savvy initiatives face the problem of a lack of reasonably priced receivers.

The webradio platforms are presented as an alternative for the salvation of radios, in particular, CRs that have faced numerous problems of operation and sustainability. Both transmission and operating problems find a solution in the streaming and webradio technologies. The example is the CR of Maxaquene, which transmits in streaming and tunable through the Garden platform.

The study allows to discover that streaming from the broadcast platforms such as Sam broadcaster or Airtime enables the transmission of radio signals to computers and other mobile devices. Basically, once connected to the Internet or not, a device is capable of playing audio files or broadcasting direct broadcast.

Realizing that CR broadcasters face technical transmission difficulties and the tendency of the new generation of listeners to move to the world of mobile and digital forum, streaming is one of the solutions to CR problems.

From there, the community of 110 CRs in Mozambique can create their own streaming platforms to compete with the urbanized radios. With this change in the way of broadcasting, which has been happening incredibly fast in the world, Mozambique can increase the presence of community webradios transmitting to global audiences.

12.4 SUMMARY

The study reflects analytically on the mutations provoked by the mobile phone on the radio from the perspective of webradio. The objective of this study is to analyze the relationship of convergence and integration of the radio in the mobile phone and its transmission over the Internet. Authors have emphasized how the CR can be saved from the technological discontinuity and from the fragilization provoked by the systems of political regulation. The focus of this chapter is CR. Some of the radio transmission technologies on the Internet have been described as paradigms to be observed when one turns to the design of new convergent and interactive languages and contents for the digital radio platform. The findings of this study conclude that radio has faced the challenge of new technologies as a factor of resignification and the mobile phone can rescue it. As a methodological procedure, this study adopts the hypothetical-deductive model, in which one starts from the perception of a gap in the scientific knowledge produced the matter and as a result prepare to formulate new hypotheses related to the complementarities of the two media, observing technological paradigms established by the mobile phone/radio.

KEYWORDS

- **broadcast**
- **community radio**
- **mediamorphosis**
- **mobile phone**
- **webradio**

REFERENCES

1. Airtime Ontario. 2018. https://www.airtime.pro/features (accessed Jan 10, 2018).
2. AMARC (Associação das Radios Comunitarian), Canada. 2018. http://www.amarc.org (accessed Jan 6, 2018).
3. Broadcast Myself. 2018. http://novastreamapps.blogspot.com/2014/07/the-broadcastmyself-website.html (accessed Jan 6, 2018).

4. Esteban, L.; Alaejos, M. The Transition from Radio to Digital Ambiance: The Experiences and the Transmigration Challenges of the Spanish Radio Industry. *Novos Olhares (São Paulo)* **2017,** *6* (2), 41–51.
5. Hamelink, C. Global Communication Policy, Rio de Janeiro. *Rev. LOGOS* **2008,** *15* (1), 10–25.
6. Hill, J. Regulatory Models for Broadcasting in Africa. *Broadcasting Policy and Practice in Africa*, 2013. https://www.article19.org/data/files/pdfs/publications/africa-broadcasting-policy.pdf (accessed Jan 3, 2018).
7. INE (Instituto nacionl de Estatística). Recensing (*Recenseamento*). In *Geral da População 2017*; Jenkins, M., Ed.; Editorial Aleph: São Paulo, 2008; p 25.
8. Jenkins, H.; Green, J.; Ford, S. *Connection Culture: Creating Value and Meaning through Propagandable Media* (*Cultura da conexão: criando valor e significado por meio da mídia propagável*); Editor Aleph: São Paulo, 2015; p 462.
9. Kemppainen, P. The Role of Public Service in Digitization of Radio. *J. Media Bus. Stud.* **2015,** *9*, 71–83.
10. Lee, R. *The Future of Pew Internet*, 2008. http://www.pewinternet.org/2008/07/01/the-future-of-the-pew-internet-project (accessed Jan 20, 2018).
11. Macedo, R. *How to Mount Web Radio (Como montar una webrádio)*, 2018. www.allsitesnet.com.br/webradio (accessed Jan 6, 2018).
12. Noticias. Eighteen Million People Hear Community Radio (*Dezoito Milhões de Pessoas Ouvem Rádio Comunitária*). *Jornal Notícias* **2016**. http://www.jornalnoticias.co.mz/index.php/ciencia-e-ambiente/63946-dezoito-milhoes-de-pessoas-ouvem-radios-comunitarias.html (accessed Jan 3, 2018).
13. Peruzzo, C. *Community Radio on the Internet: Social Empowerment of Technologies (Rádio comunitária na Internet: empoderamento social das tecnologias)*, 2006. file:///C:/Users/user/Downloads/3382-10772-1-PB.pdf (accessed Jan 23, 2018).
14. Prata, N.; Martins, H. Webradio as a Business. *Rev. Commun. Soc. (Braga)* **2011**, *20*, 129–140.
15. Radio Garden. 2018. http://radio.garden/live/maputo (accessed Jan 5, 2018).
16. SAM Broadcast. 2018. https://spacialaudio.com/sam-broadcaster-pro/ (accessed Jan 6, 2018).
17. World Bank. *Cities and Climate Change: Un-urgent Agenda*, 2015. http://siteresources.worldbank.org/INTUWM/Resources/340232-1205330656272/CitiesandClimateChange.pdf (accessed Jan 3, 2018).

CHAPTER 13

Luxury Brands on Social Media Platforms

AMIT KUMAR

ABSTRACT

The chapter focuses on the social media platform as a means of targeting potential customers in this fastest emerging community. Social media platform has expanded beyond social exchanges to business exchanges. Luxury brands too are not able to keep themselves away from the excitement of using this platform. However, does this platform really going to help? Can luxury brands cultivate expected yield? Is it the right choice for them? Similarly, many more questions have been raised by brand owners. Despite these questions, luxury brands are being placed at social media platform. This chapter provides insight into the strategic use of social media platforms for brand building and promotion. Also, it focuses on advantages that the brand owners can seek for future brand potential. Also, this chapter presents the correct perception toward the use of social platforms for brand promotion.

13.1 INTRODUCTION

Premium brands are well known for their priceless experiences and tagged social status, whereas social media platforms can be used regardless of status, class, and any racism. It is an open platform, where anyone can have social connection above their class status. For so long, luxurious brands were out of touch of these media platforms, but since last decade, great association has been observed between social media platforms and luxury brands for promotion and marketing. The ice-breaking relationship has resulted in huge publicity with almost no cost. The power of social media platforms has given them a cutting edge over various struggling brands.

This chapter unveils various issues and challenges of using social media platforms for publicity and promotion, exposes connection (reasons) of brands with social platforms, and refocuses the views of strategic thinkers on long-term use of social platforms.

13.1.1 LUXURY BRANDS AND SOCIAL MEDIA

Brands have close relationships with media since long in various forms, in printed form or in audiovisual form. However, these forms have been used for promotion, advertising, and marketing only. The world today acknowledges the presence and importance of media in best-selling the best brands. Brands are experiencing a new form of consumer behavior toward the brands, which were untouched and out of reach by majority of buyers. Electronic media platform has given a distinct place to the brands, which never thought of having physical outlets across the world. It made it possible for those brands which never dare to target global buyers due to various issues.

13.1.1.1 LUXURY BRANDS

Luxury brands have always been there in talks for their exclusiveness.[5] It always engages in creating a meaningful relationship with customers. Because, luxury brands have a sound relationship with customers due to strong brand awareness, loyalty, and equity. Luxury brands have now become very common at virtual places too, as the buyers' presence is more at the virtual market than the actual market. Popular e-tailing brands have added many popular brands in their basket to attract and retain premium buyers. Brands like *Glossy*, *Tanishq*, *Cartier*, *Burberry*, *Gucci*, *Prada*, *Boss*, and so on are doing extremely well at online shopping mart (Fig. 13.1). These brands have achieved social acceptance due to brand equity and positive consumer behavior. People follow premium brands at social media to keep up with others or meet social standards (Fig. 13.2).

13.1.1.2 SOCIAL MEDIA

Social media is quite a new platform for luxury brands. Some expensive brands are unwilling to use this modern marketing platform because they believe that brands' past performance and equity are unique selling points. The reason behind this hesitation is probably the fear of adulterating the brand image. Many luxury brands are still unsure for making their presence at social media platform because of loss of exclusiveness[5] and fear of becoming commonplace, whereas some luxury brands smell the significance of being social and connecting people through the social media platform. Gradually, luxury brands (including fashion, electronics, jewelry, hotels, etc.) are getting cutting edge with extensive customer engagement and preference over the competition.

FIGURE 13.1 Popular e-tailing brands.

Source: Reprinted with permission from www.pinterest.com.

13.1.1.3 HOW DO LUXURY BRANDS USE SOCIAL MEDIA?

Premium brands no longer keep themselves away from social presence. They have soon started fetching strategic benefits from social platforms too because the social media has rich sources of large customer pool with diversity in their demographics. Importantly, the power of electronic media (such as reach, durability of message, cost-effectiveness, and the most preferred) is customers' presence and acceptance has acquired a big buzz. Internet marketing is blessed with numerous tools for marketing and communication including corporate official websites, social media platforms, e-commerce websites, etc., but brands have found implausible touchpoints at social media. Globally, most preferred social platforms (Fig. 13.3) include *Twitter*, *Facebook*, *Instagram*, *YouTube*, *Skype*, *Hike*, and others. Some important benefits attached with social media for luxury brands are[3,8] as follows:

- *Strong brand recognition*: Brand's visibility and presence at popular social media platforms turn into a brand's voice. It actually speedups the brand to become popular and familiar to those even who do not consume the brand. It also helps in tapping new customers as well as more rememberable and familiar to the existing one. For example, people who frequently use social sites let's say Facebook, Twitter, YouTube, and others could hear about the brand accidentally first time, but from now on they will come across again and again at different networks and may find the brand more global and premium. This may stretch the brand toward global acceptance and recognition.

FIGURE 13.2 (See color insert.) Premium brands at social media.

Source: Reprinted with permission from www.pinterest.com.

FIGURE 13.3 (See color insert.) Most preferred social platforms
Source: Reprinted with permission from www.dbcinteractive.com.

- *Strong brand loyalty* is found among brands at social media platform which makes the searchers happier and attracts loyal customers. Companies have understood the importance of "being in talks" at social platforms to enhance customers' acquaintance with the brand. Strategic brand management inculcates the social act of a brand to promote and build a more trusted and loyal brand, and this is only possible when a brand is set with socially oriented goals. A study indicated that more than 50% of American customers, who are frequent users of various social sites and follow it daily, are found to be more loyal for purchasing the brands at social media. In another research conducted by Tech University, Texas, revealed that brands, which use social media platform, are more eligible for loyal customers than anything else.
- *Brand synchronization:* Social presence gives the brand a chance to synchronize with customers and can increase customer conversion rate. Brands at social platform become socialized and enjoy

strong brand positioning. Every action a brand takes at social media brings customers closer to the brand, increase visits and association. Customers always have a soft corner to the brands as these are attached to their lifestyle and offer them an attractive appearance. Brand synchronization is an effective tool for brand humanization that gives a social touch to the brand rather than a profit-making machine. A real example of brand synchronization through the social media platform is McDonald's. It has its own official twitter link, which is being followed by more than one million followers. To give a personal touch to the customers, McDonald's twitter team allow followers to tweet with their own account using their initials along with their pictures at screen. It makes followers feel that they are talking to a real person. Similarly, Walmart has millions of likes, comments, messages, and blogs from customers on its Facebook link. Through this social effort, Walmart has strong, healthy, and lasting impacts every year in the form of loyal customers.

While sharing close association, luxury brands with social media platforms experience various obstacles in different forms and formats.

13.2 VIRTUAL GAP

This section focuses on the virtual gap that lies among "social media and luxury brands." The gap is usual and very obvious because the ultimate aim of social media platforms was to make social connection while exchanging social messages among users of different geographic locations. However, the luxury brands like Chanel, Gucci, Prada, Armani, and others were known for their exclusiveness, exclusive showrooms, rareness, and others, and if these brands are easily available online where no distinctiveness will come, how they will manage brand preciousness. Despite all these, today these brands are seemingly present and available at various social media platforms. Therefore, the current study found the gap: why do these brands use social media platforms? Is it fulfill their brand's objectives? Will it maintain the exclusiveness and rareness of the brand?

The study is based on the research question: Is it the right choice to be present at social media platforms being a luxurious or exclusive brand?

13.3 VIRTUAL GAP: METHODOLOGY

The study describes the fallen truth behind the motives of the presence of luxurious brands at various social platforms for getting awareness, popularity, and preference. The study examines and follow popular social platforms in India and monitor the brands' presence. It concentrates and consider the comments, suggestions, likes, or dislikes for these brands by the users. The study also accumulates and compiles the valuable findings and outputs of the previous searches on similar area and topics. As the study follows the "systematic review" technique, it focuses on literature evidence on similar and the related topics to draw another aspect of understanding and meaning on the research question. At the end of the study, readers may come to know about various untouched facts (reasons) of social media platforms and luxury brands' presence.

13.4 GAPS IN SOCIAL MEDIA PLATFORM FOR LUXURY BRANDS

Social media undoubtedly is a great platform for branding and offer various strategic advantages, but it is not complete in its own. It has various limitations and gaps for brands in their promotional campaigns. Luxury brands available on social media platforms always encounter difficulties in presenting brand's worth and communicating brand value. Significant advantages are as follows: a luxury brand carries cannot be displayed on social platforms due to media limitations and the brand's own strategic intent. One of the biggest gaps in social media for promoting the brand is brand's inability to communicate its standard, uniqueness, luxury, and so on. Short messages on social media cannot create a buzz as desired by the brand. In the mass media age, every brand is using social platform for brand promotion along with conventional techniques (viz., short storytelling, cinematic tricks, storytelling, etc.) and has found that this medium is half done. Short messages do not fulfill the brand's desires, whereas long messages are disliked by customers. Under such situations, it is hard to get a grip on customers with open promotion sources, where people are left with an open choice to give attention to the campaign or not.

Any media message is considered worthy when it can attract target groups with lasting effect. Social media as a source of innovative marketing technique derived from high technological can change across the world and functions have to significantly redesign people's communication style and interaction on the Internet. It becomes a challenge to be attractive for a brand, where social websites become challengers too in gripping people and hold on at a webpage. Though it is easy to make presence on social media, yet it is tough to be attractive. Personal brands' messages racing for customers' visits and clicks to achieve high traffic may earn favoritism. Numerous luxury brands have supported the thinking that it is the biggest gap in social media to show oneself attractive and different.

Luxury brands win over the market with its premium image and exclusivity. In retailing, brands are undisputedly inimitable as it is sourced with luxurious infrastructure and other tangible attributes. It is easy to show the exclusivity of the brand. It is supported by end number of features including well-recruited human resources. However, it is an equally cumbersome process when it comes to social media usage. Showing brands' exclusiveness online is always a challenge to the brand manager. With the lack of stuff necessary to spread, luxury online media platform has a big gap for brands presence online.

Media management: Social sites cannot manage messages from brands available on their web page. All brands pin at same page competing each other. Therefore, it becomes difficult for the people visiting social sites to experience brand's luxury to differentiate one another. Popular and unpopular brands use the same platform for promoting their brand. There is a big gap in such platforms, which do not allow premium brands to show their uniqueness differently. Putting all brands into the same basket bring homogony thus reducing brand equity. At social media, only those brands achieve customer preference which are known to buyers and the rest all get casual approach and view.

13.5 IS SOCIAL MEDIA PLATFORM A RIGHT CHOICE?

Luxury brands are overwhelming in getting consumers' response offline.[6] However, for a subsequent growth of the brand, it is advisable to get social buzz too. Brands at social platforms are experiencing constructive engagement of buyers for further brand proliferation. Ideal success for a brand is acquired from being social with customers at social platform and frosting through social engagement. There are numerous examples, where premium brands are materializing their efforts at social platforms to meet business goals (Fig. 13.4). Collaboration of online and offline media for brand promotion has brought dual opportunity to explore market, where social presence is seeding grand awareness among the people and word-of-mouth publicity bringing financial prosperity for the brand.

Examples: Brands Using Social Platform

**TOPSHOP: A blasting fashion brand used "Twitter" during London Fashion Week 2015 to approach youths for promoting the brand with exciting colors, shapes, and glamor. The brand engaged more than 3.8 million customers in its campaign named "LIVETRENDS." Sales were increased by 75% on various featured products.

**Dior J'Adore: A dominant fragrance brand used "YouTube" for its 90-s ad commercial to get quick viewers buzz. The brand is known for its fabulous fragrance, glamorous femininity and sophistication. This footstep caused 1.1% increase in market share, 17% increase in sales, boosted brand name throughout the world.

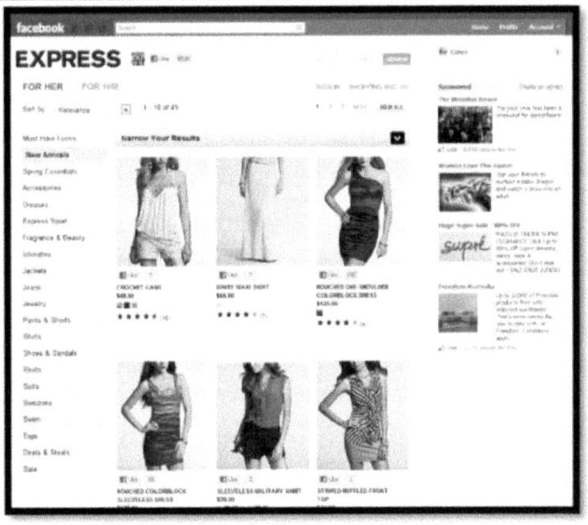

Women's and men's clothing at <express.com>.

FIGURE 13.4 (See color insert.) Why social media is important?

Source: Reprinted with permission from www.jeffbullas.com.

A renowned sports shoe brand "Nike" launched its campaign to connect people with the brand more closely and emotionally. In its campaign named "NIKEPHOTOiD" created a burst among the customers at "Instagram." Over a short span of time, the brand achieved unmeasurable popularity and awareness with more than 10 million followers. Customers sent various photos with Nike shoe, which helped the brand in getting a personal touch with customers. With a grand success of this campaign, Nike in 2013 designed a microsite for customers to allow them to customize their runners according to their Instagram pictures. This effort made customers' cocreation popular overnight and was shared across almost all social media platforms. Nike with almost no media expenses achieved higher rating, millions of self-designed shoes with customer satisfaction and never estimated clicks and visits on the official website of Nike.

Social media platforms are one of the best methods of reaching customers and promoting the brand with controlled cost in today's scientific world. There are numerous examples of premium brands using social media to create hike for their brand. Marketers believe that such platforms grab considerably more customers' engagement than any other. It is because social media platforms provide ease of putting messages along with images, which is more eye-catchy than simple word messages. Tweeting with images always gets preference of people because it is more descriptive, colorful and informative, similarly, a photo ad on Facebook gets more customer engagement and sharing.

13.6 WHY IS SOCIAL MEDIA?

Commercial use of social media[1,6,9] provides a competitive edge to the brand along with various strategic advantages as described below:

- *Brand humanization*[2] is based on the fact that the brands with social presence are more likely to be live to the buyer. It interacts with visitors and behaves in a humanitarian manner to jell with people and converting them as long-term prospects of the brand.
- *Build brand equity* is always an asset to the brand. Luxury brands are building brand equity by connecting people personally at social platforms.
- *Concrete loyalty:* Social presence of luxury brands undoubtedly concretes customer loyalty in the long run.
- *Connect long-term audience:* Social interaction creates chances of connecting people for a long time. Luxury brands keep customers informed to retain and sustain brand performance.
- *Customization:* Social platforms allow customization of messages to give a personal touch to the customers. Customized messages appropriately target customers and increase customer conversion rate.

- *Public relations:* Brands through social platforms create long-term relations with customers and followers. Official websites get likes, comments, suggestions, and many more.
- *Relationship capital formation:* Social platforms have been proved as a strategic place for relationship capital formation, which provides an opportunity to explore the wider market through word-of-mouth publicity. Relationship capital gives brand preference and recognition in the long run.
- *Social sync:* Being social shows the social involvement of brand to viewers. It reveals how brands are socially synced.

13.7 WHAT IS "EXCLUSIVENESS"?

In the high technological world where information dissemination is easy through cross channels, multimedia and customers are well informed with plethoric information; therefore, the ultimate question is: Why you? Every premium brand has to answer this question with logic and at the real ground. The brand has to show its exclusiveness as a point of attraction to the target group because affluent customers not only buy a premium product but also purchase the lifestyle a brand portrays. Exclusivity reflects from a brand in different manners like some brands have price exclusiveness, customers buy these brands to live a luxurious lifestyle and get satisfaction to pay high-end price for the brand, whereas some have in the form of "Best, First or Only." Customers choose that brand due to limited series or model. Many premium brands have service exclusivity with them through which they target high end customers. These brands express their exclusiveness in their services during sales, during consumption, and after sales. Brands add unique product range with none other than models to keep the brand fresh and exclusive for the customers, it also helps them to allure customers' preference.

Consumers look for the product, which is uncommon and limited in quantity to experience luxury and to maintain exclusiveness, whereas brands maintain exclusivity by releasing exclusive products. It reveals the expertise of the manufacturer or designer for its creation. For example, luxury watch brands design a few pieces of watches by keeping customers' desire to own a rare one.

Jaeger-LeCoultre (Premium Wrist Watch Brand), as a part of its brand strategy, celebrated the 180th anniversary of the brand by introducing a limited edition three-piece collection named "Jubilee." The limited edition series promoted during the "Venice Film Festival" to build customers' concern in the world's renowned watchmaker's brand history. This stunt made customers aware of the worth of the brand, how old and precious the brand is and narrated the successful journey of the brand.

A well-named French Glass Maker "Lalique" joint hand with very renowned Swiss Watchmaker brand "Parmiagiani Fleurier" to parade their expertise and craftsmanship by creating and promoting a limited edition table clock. The clock was so unique in its own due to limited in quantity, the watchmaker can produce only five in quantity in a year because of labor-intensive requirement by the product. This innovation in product line created hype in the market and brought huge popularity for both the brands.

13.8 SOCIAL MEDIA PLATFORM: LUXURY AND EXCLUSIVENESS

Being an open and freely accessible platform, social media platform always puts a challenge before luxury brands to maintain and display brand exclusiveness. Brands have experienced numerous pitfalls to distinguish themselves from other brands. It has become a terrific job to target the right audience and community for the brand. In this new digital age, social media platforms are being treated as a new market place, where target groups are having digital natives to share information, communicating brand preference, displaying the favorite brands, and talking on to popular brands. This new generation of customers has a different approach to the brands to be selected and preference made through digital response and presence of the brand. Premium brands' presence on social media attracts new prospects along with the customers, who expect their loveable luxury brands to use social network to share interesting contents added. Brands use various techniques to display how exclusive the brand is, as compared to other competitive brands in the market. Some of the workable strategies for maintaining exclusiveness online are as follows:

- *Brand associations:* Customers weigh brand's worth through its associates: who are the corporate partners? With whom are brands partnered? Brand ambassador organizes the events for the brands and approaches to different media, and so on. It shows how exclusive is the brand because customers feel the heights of luxury or exclusiveness while using the brands. The high-profile association grants the brand a high-profile image and class, which enable the brand to show its luxury and exclusiveness.

- *Standing or character of the brand:* Brand's standing in customers' psych resembles in their talks on social media platforms. Many luxury brands in the world restrict their presence on social sites just to maintain their exclusiveness as they believe that being social become being common. However, this phenomenon is quite unreasonable in the digital age of marketing. Customers talk about brands, irrespective the brand is present or not at social platform; discussions are already happening and will continue to happen regardless of what is the strategy of the brand for media engagement. The brand's reputation is being measured also through social media engagement and its global reach. Global brands have global customers and characterized by their social function there only, where the customers belong to. For example, if the brand has its customer base in China it has to be present at various Chinese social platforms to engage in social talks of the brand, following customers' conversation about the brand and to connect people outrightly through social media, which provides global coverage with comfortable language capabilities.
- *Channel through popular blogs:* Premium brands use famous blogs to share information and contents with target groups with an aim to catch exclusive customers while showing exclusiveness of brand. Exclusive blogs are followed by like-minded customer groups that respond in the same manner and behave likely. Therefore, luxury brands channel these blogs for information sharing and showing brands' worth.
- *Display luxury lifestyle with the brand:* High-class customer segment looks for unique brands to represent their social class. Brands, which prove their credibility to offer high-class experience with brands, get more satisfaction and sophisticated customer base. Therefore, brands at social media platform portray a luxury lifestyle around the brand. For example, a 7-star high-class hotel makes its social presence with displaying its ability to cater all of customer's requirements with luxurious amenities. It carries all premium brands to serve customers and show how the brand creates luxury for customers around it.
- *Storytelling:* Luxury brands share success stories of their brands, where it stresses on the luxurious experience of customers with

glamor, passion, class, and much more. Popular brand's history is narrated in short blogs particularly pointing brand's engagement in creating high-class values for the customers. These stories create an interactive effect on customer's psych for brand formulation and awareness.

13.9 WHY ARE LUXURY BRANDS BEING SOCIAL?

Socialization of luxury brands is a common practice among brands, as marketers presume that to measure the brand's effectiveness and performance social metrics (including likes, sharing, tagging, comment, etc.) is also an important tool. Brands, which rank top on social platforms, get unmeasurable awareness with a distinct image in customers' psyche. It is believed that social media is ranked among the best practices available in today's digital world for successful brand positioning. Therefore, companies' investments into digital media including social media is increasing undisputedly. Being social is considered as the smartest way to reach target groups strategically by minimizing expense and optimizing return, that is how luxury brands take social media.

In today's market scenario, people have social concerns and share their views with social groups, talking about their favorites, expressing comments, and so on.[6] Hence, companies found numerous reasons to be socially active on social media for not losing a single change of acquiring and making customer base (Figs. 13.5 and 13.6).

- *Big apparel brands* use social media connection to distinguish marketing strategy, where these brands connect customers to promote customer cocreation. It not only provides customers' innovative ideas for product innovation but also make customers feel part of that brand. Such brands accept customers' ideas about how they want to see their brands? What innovation they would like to make into? What features would they prefer in the existing product? How to customize the brand for specific customer group? and so on. Such efforts help brands to connect customers emotionally, socially as well as psychologically.
- *Brand socialization* refers to brand access to all categories of society, and social media platforms provide customer access and help the

brands to be socially connected with different groups of society. Brand socialization is an effective customer control mechanism.

FIGURE 13.5 **(See color insert.)** Specialties of Adova Soft.

Source: Reprinted with permission from adovasoft.com.

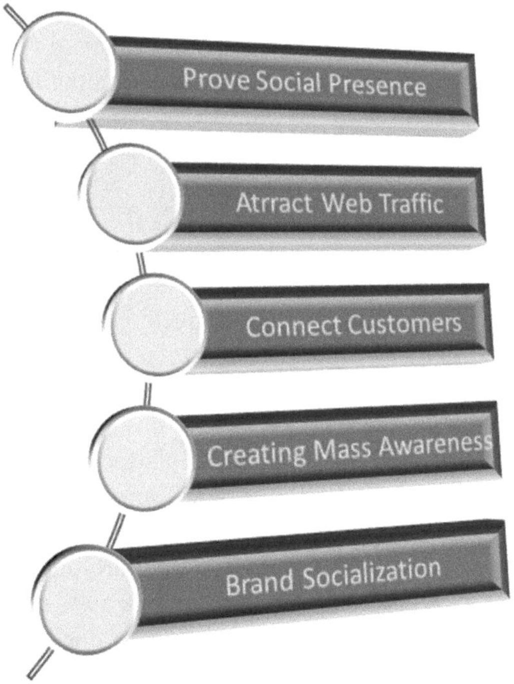

FIGURE 13.6 **(See color insert.)** Potential reasons for using social platforms.

- *Brands* like Samsung, Apple, Panasonic are working on generating followers online. These followers become an integral part of brands for formulating and implementing strategies to attract more viewers. Brands keep on innovating their website and social web page to attract more traffic (Fig. 13.5). More the traffic, more the chance of increasing the customer base. It is believed that more than 80% of followers are the buyers of that brand or intend to buy. By putting such efforts, brands ensure signaling to new entrants for value creation to each other.
- *Consumers* believe that the brands are socially active and are more trustworthy. Every 7th out of 10 customers know the brands that are being promoted at social media platforms. Brands have understood the psyche of customers in terms of trust and reliability of a brand; therefore, they try to prove their social presence at large.
- *Mass marketing* is a mantra of success in today's business world; mass marketing in digital world has become the most cost-effective method of promotion to offer relatively better results than any other traditional source. Through social platform, brands attempt for mass marketing to connect nonbuyers also. The strategy behind this is to prepare nonbuyers into buyers in the near future through creating mass awareness.

13.10 BEING SOCIAL OR REMAIN OUT OF TOUCH

Superior brands are very keen to know what should they do to get a clear picture of "Being social or remain out of touch"? It is believed that premium brands offer the highest value to the customers and these should remain in scarce. Because high-end customers buy brands, which are in limited edition or superior. However, still out of discussion, luxury brands are targeting customers through social media platforms.[4] Competition led the introduction of technology and technology spread ease of doing business with customers are living the technology at every sphere of their life; therefore, marketers have identified e-sources that are best ways to reach out customers. Various societal changes have come up to redefine the market, where customers look for product and services that are easily available. Social media platforms provided what actually the mass customers were looking for. The current demand of the market is to be alive electronically on different platforms to capitalize on the opportunities. This sounds great to luxury brands that such premium brands enjoy inherent advantages while delivering values to the customers using social media platforms. Brands like Cartier and Chanel are getting customers' positive response and preferences on LinkedIn than ever before. Somehow it connects potential customers to brands socially and emotionally.

Over a period of time, technology in marketing has brought various optimistic changes in finding, converting, and sustaining customers. Online social websites are among them to accumulate customers' strength at a single platform. It offers various strategic advantages to the brands present online, such as large pool of customer base with continuous interaction on enhancing brand performance (quality, features, performance, and so on), wealth accumulation in the form of adding new users' advice by existing one, followers who like and comment on customers' write-ups for brands etc. Luxury brands do not provide only luxury but also an experience, which can never be obtained from any other sources. Various market researchers have convinced from the strategic outcomes of social media promotion as it is the only platform where marketers can find customers with free time and in relaxing mode.[7] People surfing on social media are more in a convincing situation, accessible, give proper attention to displaying brands and participate in conversations.

Another attractive angle of being social is the global cost-effectiveness of the luxury brand promotion. Premium brands represent the global image, hence they need to spend heavy investments over brand promotion globally as it may increase the cost of operation resulting in shrinking profit margins. However, these brands are now channeling through social platforms to control cost and make a global presence. Social media platforms give free hand to the brands for accessing the global market and limiting advertising and promotional budgets. It is the best way to build communities for global brands.

13.11 CHALLENGES FOR LUXURY BRANDS ON SOCIAL MEDIA PLATFORM

13.11.1 NONADOPTABILITY OF POPULAR SELLING TECHNIQUES

Luxury brands are found detached from popular selling techniques as the brands target elite customers. Techniques like discounted pricing, BOGOF, bulk selling, etc. are nonworking methods in case of luxury brand selling.

Marketers have identified the fact that these techniques put an adverse effect on affluent customers. Price discount and offers do not work for luxury brand promotion as these brands are being purchased to maintain social image, luxury, and uniqueness. Across the globe, various luxury brands are targeting customers with pricing strategies just to maintain present customer base as well as to attract new ones. Managing brand equity at social media platform is again the biggest challenge before luxury brands because luxury means not only to maintain class but also to create an identity as oneself in the society as well.

13.11.2 COUNTRY-SPECIFIC BRAND CROWD

Social media is a global platform, where customers from different countries are found active and talking with each other. Therefore, the platform is flooded with various types of country-specific luxury brands along with globally accepted brands. These country-origin brands create indirect competition to the known luxury brands with people's recommendation to their social friends. Sometimes, customers at social media have found the best alternative of their favorite brand at a competitive price with all-inclusive. It happens due to cost-effectiveness of a social media platform for brand promotion. Small brands too use the same platform for promotion, which has become the biggest challenge to maintain differentiation.

13.11.3 FEAR OF GETTING COPIED

The social presence of luxury brands with a description of features and attributes may create a parallel competition through imitation. Most popular brands are always under threat of getting copied at various locations of the globe through social interface. This parallel market structure creates customer confusion, doubt on the reliability of the product, brand become commonplace, and so on. Hence, brand has to make a continuous effort to prove its genuineness over social platform.

Value creation: Popular luxury brands are the highest value creators for ultimate consumers in many forms. They communicate value to customers through various media ads and campaigns, but when it comes to the social media campaign, it seems quite difficult to express brand value in terms of comfort, luxury, uniqueness, elite, and others. On the other hand, the type of value a luxury brand offers does not relate to a larger group of customers as they are not the right audience. Common social media platforms are rich sources of customers but not for highest luxurious brands, whereas elite social platforms are Twitter, LinkedIn, etc.

13.12 ONLINE VERSUS OFFLINE SOCIAL PLATFORM FOR LUXURY BRANDS

Among all methods of promotion, online media mode is unique and exclusive for several reasons. However, it still has various limitations for luxury and premium brands, which make a difference between online and offline media. Some of the limitations, which are inherited with online social media platforms, are discussed below:

- *Failure to communicate the brand's standard:* Social presence of luxury brands sometimes incur huge awareness but loose distinctiveness. It is quite difficult to show the brand's standard or how sophisticated the brand is online? Social platforms do not provide the opportunity to show exclusiveness; therefore, luxury brands fail to show its level of standard. Premium brands available online always adopt a unique way to deliver its message.
- *Fear of being common:* Commercial ads of luxury brands go across people and groups in a repetitive manner. This repetitive nature of ad commercial on social media makes brands common to all, which somehow affect the uniqueness of the brand. Repeatedly sharing the same message by people make them feel that the brand is struggling for customer acquisition and positioning.
- *Hard to measure results:* Brands are investing valuable resources (human resource and time) for their presence at social platforms. Media presence always cost a lot to the brands, but it is being used for successful brand creation. Each media spends and their return can be measured and can estimate the effectiveness. However, measuring social media effectiveness and return to the brand is quite a difficult task because there is no measurability scale or criteria. Messages can only be counted through likes counting, visits counting, and others but still, sometimes it will not help the brand in the exact estimation.
- *Loss of exclusiveness:* Brands are called premium or luxury, which maintains scarcity and exclusivity in its sale. Brands that are found

rarely among few customers may earn unmatched image with the symbol of exclusiveness. Social media platform is an open-access platform, where people of all classes are wondering. When a luxury brand goes through social media, it becomes popular and is known to everyone. Many times such stunts cost a lot to the brand and premium brand becomes common. This is a major threat to the brands which are premium and offers luxury to the customers whereas offline method uses a specific route to approach to the right target audience.

- *Managing messages:* One of the biggest challenges of social media platform for luxury brands is to manage incoming messages that are being shared openly on fans page or on wall. Many times, brands also receive negative messages shared among people and written on social websites. This may bring negative perception of customers and may lead to negative comments. Message management is a challenging task in social media marketing; therefore, brands are under pressure to optimize their use of social media for brand promotion.

13.13 SUMMARY

The chapter focuses on clever use of social sites to spark buyers' passion about premium brands to choose and advise. Selection of right media platform is the biggest question before these brands; however, the chapter may help to make decisions about how to choose media, what parameters should be considered while playing with social platforms and how can cultivate more out of it.

KEYWORDS

- **brand communication power**
- **brand's social power**
- **interactive platform**
- **premium brands**
- **social media platform**
- **social media transformation**

REFERENCES

1. Boyer, N. *Get Social: 9 Ways to Promote Your Content and Brand on Social Media*, 2015. http://www.business2community.com/social-media/get-social-9-ways-promote-content-brand-social-media-01411863 (accessed Jun 30, 2018).
2. DeMers, J. *Why It's Important to Humanize Your Brand on Social Media*, 2013. http://www.business2community.com/social-media/why-its-important-to-humanize-your-brand-on-social-media-0445512#!by85az#x0ZbzaeQe6vLO6DT.97 (accessed June 30, 2018).
3. DeMers, J. *The Top 10 Benefits of Social Media Marketing*, 2014. http://www.forbes.com/sites/jaysondemers/2014/08/11/the-top-10-benefits-of-social-media-marketing/#3f4732a4d2c2 (accessed Jun 30, 2018).
4. Edgeworth, A. *Top Social Insight for Targeting Canadian Millennials*, 2018. https://www.synthesio.com/blog/targeting-canadian-millennials/_(accessed Jun 30, 2018).
5. King, J. *Exclusive Products Help Bolster Timepiece Brand Mythology*, 2013. https://www.luxurydaily.com/exclusive-products-aim-to-support-luxury-allure/ (accessed Jun 30, 2018).
6. Meredith, J. *Six Ways for Luxury Brands to Maintain their Allure with Social Media*, 2013. http://www.synthesio.com/blog/6-ways-for-luxury-brands-to-maintain-their-allure-with-social-media/_(accessed Jun 30, 2018).
7. Shahjahan, A. T. M.; Chisty, K. U. Social Media Research and Its Effect on Our Society. *Int. J. Inf. Commun. Eng.* **2014,** *8* (6), 2009–2014.
8. Standberry, S. *75 Benefits of Social Media Marketing*, 2016. https://www.lyfemarketing.com/roi-benefits-social-media-marketing/ (accessed Jun 30, 2018).
9. Whiting, A.; Williams, D. Why People Use Social Media: Uses and Gratifications Approach. *Qual. Mark. Res.: Int. J.* **2013,** *16* (4), 362–369.

CHAPTER 14

Role of Artificial Intelligence in Sustainable Development

MOUNIA KHELFAOUI

ABSTRACT

For more than a century, man has transferred his intelligence to machines to create a behavior that is close to his own. It is a complex system that implements the knowledge of the unconscious sciences that describe human, animal, or artificial thought. To do this, it uses sophisticated electronic processes and powerful computers. Artificial intelligence (AI) tries to imitate perception, reasoning, and understanding of events. It has become a global paradigm forcing scientific communities to discuss the utility of AI in our lives. Through the opportunities offered by AI, the human being assumes the well-being that is realized in three aspects: social, economic and environmental. Like all human activity, it has aroused the scientific curiosity of researchers about the quality and impact of their externalities. These ideas are very close to the concerns raised by the definition of sustainable development. Thus, the purpose of this chapter is to define the link between AI and sustainable development. The author seeks to determine the contribution of AI to the concretization of the objectives of sustainable development in our lives.

14.1 INTRODUCTION

Humanity has taken a leap forward in information technology and artificial intelligence (AI). The latter offers opportunities for users to excel in creativity. The feats of the AI are numerous and of very satisfactory qualities to the point, where the machine intervenes in processes that were reserved exclusively to the man especially in treatment of the images and the videos, treatment of the language, the predictive analyzes the programming of the games, intelligent automation, and the imitation of complex systems.

Technological breakthroughs are paving the way for researchers to embark on in the field of AI. In a few years, the machines will become conscious to manage the daily life without the man being able to have control.

Like various human activities, economic activity has seized this technique to take advantage of its attributes. As a result, more and more companies are using the services of the AI to enhance their business and earn more money.[24] Indeed, AI has an apparent impact on economic activity. It can change the economic theory, for example, it can influence theories such as the decision-making theory or the efficient market hypothesis or the theory of perspectives.[20] Its objective is to make the economic agent more rational to perceive the world around it and to take proper decisions. "*Put another way AI researchers aim to construct a synthetichomo economicus, the mythical perfectly rational agent of neoclassical economics.*"[28]

In this case, the question challenges us: practically, how does AI influence consumption, investment, employment, and so on. In response to this question, the Price water house Coopers (PWC) report of 2018 argues that research on the impact of AI, about the work of Frey et al.[11] and Autor,[2] focuses on the effects of AI on jobs that will be replaced by automation.[32]

Practically, AI affects several variables of economic activity, not just jobs. Companies have turned to the digital age by having recourse to the techniques that offer them the AI. Indeed, the ability of information technology to process data (collection, storage, and analysis) must permit companies to improve the quality of their products and reduce purchase delays, which aims to increase demand.[14]

The impact of AI has also marked the world of finance. Thus, the issuance of money in industrialized countries has grown exponentially. The movement of capital and cash flow has accelerated particularly toward emerging countries.[9]

Concerning economic growth resulting from AI, the analysis by researchers, estimate that "Global GDP will be up to 14% higher in 2030 as a result of the accelerating development and take-up of AI (the equivalent of an additional $15.7 trillion)."[33]

However, opinions are not all optimistic about the role of AI in economic growth and do not see the same feats of this technology in this area. Some reactions question the place of man in the spectrum of automation, the welfare of the population, business productivity, and the revival of the economy.[13]

The rationality, efficiency, or profitability required by economic activity are no longer sufficient to ensure the welfare of man. This situation has been denounced several times. It gave birth to a new vision of economic growth or sustainable development (or sustainable growth) that meets the needs of current generations without forgetting future generations.

As well, does the economic growth supported by the IA consider the well-being of future generations, does it control the externalities of the economic activity and their impacts on man and the environment natural.

The purpose of this study, in this chapter, is to determine the link between the two concepts: SD and AI. The author tried to determine the role that AI can play in achieving the goals of the third millennium brought by the SD. Would contribution of AI in SD always be beneficial to humanity? Alternatively, will it need a global summit to say or "Stop AI?"

The second section is devoted to the overview of AI and short history. The third section deals with new aspects of economic growth in the context of the digitization of economic activity. The concept of SD has been analyzed in the fourth section including its genesis, its objectives, and means by which it will be introduced, and its sustainability. The theme of the interaction between the AI and the SD is current and it is in full conceptualization. The fifth section fits in this line of ideas and tries to determine the interactivity between the two concepts.

14.2 REVIEW ON AI

AI aim to imitate the reasoning of the human brain in decision-making. The simulation of human intelligence is executed through high-tech digital machines especially computers. AI can involve all areas of human life with many applications. Without being exhaustive, we can mention expert systems, language processing and problem-solving, knowledge representation, robotics, learning, and complex systems.

In the expression "Artificial Intelligence," we associate two words: intelligence and artificial. Etymologically, the word intelligence designates the appetite to make the right choice. The word "artificial" means an object realized in the rules of art.[3]

Thus, the etymological meaning of this notion inspires the elaboration of its definition with some reservations. It does not know a consensus because of the difficulty to define the terms "intelligence" and "artificial," the dictionary alone gives several definitions.[16]

However, there are several definitions of the intellectual intelligence aimed at defining the purpose that a system must show to enter the category. Either, analytics, which means that they are deploying a list of capabilities required to create an AI.[23]

Several definitions listed in the literature are as follows[21]:

- "The study of mental faculties using calculative type models" according to Charniak and McDermott.[7] They insist on the behavior of the human being in the system;
- "Intelligent agent design" according to Poole et al.[29] This definition focuses on the role of the machine in the system;
- "Discipline studying the possibility of having the computer perform tasks for which man is better than the machine today" according to Rich and Knight[38] and "The automation of activities associated with human reasoning, such as decision-making, problem solving, learning" by Bellman.[4] The authors consider that the operation of the system that matters;
- "The study of the mechanisms allowing an agent to perceive, reason, and act" by Winston.[39] According to this vision, the most important task is allocated to the machine;
- "Studies of entities with intelligent behavior" by Nilsson.[26] This definition is purely material, it incurs on the machine in the system.

Therefore, AI is none other than human intelligence transferred to a machine. Its purpose is the creation of a system; whose decision-making behavior is close to that of the human being. In other words, and for the sake of comfort, the man has delegated certain tasks that he considers cumbersome to machines.

John McCarthy coined the term "IA" in 1956 at the first conference on this subject.[8] Only, the idea of AI goes back some years earlier. During the period 1943–1955, work considered as the beginnings of AI (the term did not exist yet), according to McCulloch and Pitts, Hebb, Minsky, and

Edwards. The latter set up the first network of artificial neurons.[22] The IA concept was concretized by the ingenious test of Turing in 1950. It is a proposal of test of AI based on the faculty of a machine to imitate the human conversation.

The period 1952–1969 flourished for the field of AI. It was the revolution of cognitive modeling when many programs were developed to solve problems. The period 1952–1969 flourished for the field of AI. It was the revolution of cognitive modeling. Many programs were designed to solve problems. Samuel's Sheckers program, Newell and Simon Logic Theorist, GPS (General Problem Solver), and Gelertner (Geometry Theorem Prover) elaborated to confirm the mathematical theorems. The rise of inventions continued with the McCarthy program named LISP (LIst Processor). It was developed for computers in order that the system could use common sense in the execution of its instructions.[1]

During the period 1965–1970, AI had a phase of decline.[25] Researchers failed in machine translation. They were very optimistic in their forecasts. The translation they developed was purely syntactical. It did not consider the semantics of expressions. The disappointment was such that Minsky and Papert published a book entitled "Perceptrons, 1973" in which they demonstrated the failure of the networks of neurons of that time.

The years of AI regression did not prevent research from hitting and overcoming obstacles. Indeed, the expert systems appeared between 1969 and 1979. The AI made its entry into the industry in the 1980s. Companies took advantage of this technique and created their own research groups. The AI then entered the era of industrialization that required the financing of major projects. In the mid-1980s, neural networks made their appearance with great success thanks to the rule "Back-Propagation."

Now AI has developed into a separate discipline. Research in this area has specialized and a modern approach has been developed to classify it into two categories: (1) the first classification is the general artificial intelligence (GAI) that is interested in the founding question of the AI—is it possible to create a machine capable of adopting the human performance and (2) the second classification is called the narrow or weak AI. It is not concerned by the creation of an intelligent machine. She deals with the resolution of specific problems.[12]

These new approaches have made possible to assimilate AI in all fields, particularly in economics, since its application no longer involves any financial risk.

14.3 ECONOMIC GROWTH IN THE ERA OF THE DIGITIZATION OF ECONOMIC ACTIVITY

Man's technological prowess continues to grow; they have transformed our lives on all fronts. For two centuries, he sought to make his life easier by inventing electricity, steam engines, microprocessor, and many other tools. Its ingenuity leads him to the creation of machines that can compete with intelligence through research in AI. The man gave the automated system several tasks that were inaccessible.[15]

Like other human activities, the economic activity is jostled in the digitization. Since then, several questions challenge thinkers in economics. What are the effects of AI on economic growth? What about the factors of production? What will be the impact on business and market structure?

The answers to these questions are two-fold: optimistic and pessimistic. Certainly, there is a remarkable decline in the evolution of investment in capital and labor to propel economic progress (capital and labor being the driving forces of production), but the doubt that animates some is not justified.[30]

The 2016 Accenture study of 12 developed economies indicated that AI has become a new factor of production with the potential to double the economic growth of these countries in 2035. It overcomes the physical limits of capital and labor and opens new sources of value and growth.

AI will accelerate growth in all sectors, especially those with strong potential. Transport and logistics, supply chain and retail trade[18]:

- Due to the intelligent programming of the traffic, we will save time in the transport. Soon, vehicles will be autonomous with an improved driver assistance system;
- The AI allows the management of the huge volume of data. This will be used in demand planning, product specification, pricing, optimization of workflow, automation of operations, and traceability;
- The contributions of the AI are the customization of design and production, the management of inventories and deliveries and the anticipation of demand and customer orders.

Forecasts of global GDP are very promising with AI. Thus, "the biggest economic gains of the AI will be in China (26% of the GDP in 2030) and in North America (14.5%), for a total of $10.7 trillion and nearly 70% of the global economic impact."[34] On the other hand, the consequences of AI on the economies of developing countries will not be as successful because of the low adoption of new technologies by these countries.

According to the Accenture study, the sectors of information and communication, manufacturing and financial services will rise in the growth rate of gross value added. They will reach 4.8%, 4.4%, and 4.3%, respectively by 2035, representing an additional increase of $6 trillion in all the three sectors. The same is true for the social services sectors and education; their benefits will grow from 109 to 216 billion dollars.[5]

Among the highlights of the AI that must play for the workforce is the improvement of public health. Indeed, the powerful theology of AI can predict, where the epidemics will happen. This will help anticipate care and avoid losses in the population.[10]

The researchers' response to the reluctance of some to the loss of employment due to AI is reassuring: "*AI will not replace work (including capital), but it will allow it to be used more effectively.*"[31] However, the technology will open for other related business with the AI. The data analyst, data processing consultant, and responsible for information security are already popular and much sought after by companies.

Facing this enthusiasm for the benefits of AI, voices rise to denounce negative impacts related to the adoption of this new paradigm. They do not appreciate the momentum of economic growth that does not consider the negative effects of economic activity on man and nature. These reflections stem, from an approach of durability that believes that economic development should not neglect the "become of future generations."

14.4 AI IN THE CONTEXT OF SUSTAINABLE DEVELOPMENT

"Sustainable development," this term conquered languages in the 1980s after its appearance in the Brundtland[1] report following the meeting initiated by the United Nations. Its aim was to propose a global program of change in development concepts and practices. This report showed the urgency of rethinking our ways of living and governing. The environmental problem was not taken into consideration until the 1960s it was about isolated behavior to fight against the different kinds of pollution. Priority was given

[1]This report is also known as "Our Common Future."

to industrialization to the detriment of the degradation of the atmosphere, agriculture, natural resources, and the welfare of the population.

The 1970s saw a turnaround of the situation. Environmental issues were taken into consideration and debates on the protection of the environment follow one another. They dismiss the negative externalities of industrialization that cause enormous ecological disasters and are widely publicized to raise public awareness.[19]

Environmentalists criticized industrialization for the momentum of growth that led to the depletion of natural resources, in quantity and quality, but also its impact on the economic and social components. The fruit of this excitement is the 1972 Stockholm Conference,[2] "the United Nations Conference on the Human Environment."

This conference brought reflections on sustainable development, which was taken up by the Brundtland Commission later. In 1980, the concept of "Sustainable Development" came into being and gained international recognition after the 1987 Brundtland Report.

Sustainable Development" is defined as *"development that meets the needs of present generations without compromising the ability of future generations to meet their own needs."* The principles of sustainable development are four-dimensional, namely[17]:

- Constitute it,
- Economic efficiency,
- Social equity,
- Preservation of the environment and participative democracy.

Thenceforth comes the practical application of sustainable development through actions such as: combating different types of pollution, reducing poverty, rethinking new methods of growth, widening the niche of economic actors, and many other initiatives in favor a global development respectful of man and nature.

Sustainable development is at the onslaught of any change in economic activity susceptible to cause damage to human well-being and to nature. Thus, the advent of AI has rekindled the concerns of ecologists about the possible negative effects of automation of the economy.

---

[2]Conference held following the publication of the Meadows Report, which stated that humanity could not continue to grow at this fast pace; it must slow down its momentum. It brought together 113 countries including developing countries and communist countries.

The debate by economists on pro-sustainable development focuses on the impacts of AI on our lives. They apprehend the potential consequences of AI on the social component.

Their fear is legitimate because over the past few decades unemployment seems to have increased. With each successive economic cycle, the troughs of the unemployment graph move upwards. Some economists believe that we are already in the early stages of a critical period when large-scale unemployment due to automation is inevitable.[27]

For developed economies (the United States, EU, and Japan), the AI is an advantage to compensate for the loss of productivity observed since the global financial crisis. Nevertheless, disruptions can occur on the job market. In fact, a PWC study found that 37% of workers fear for their jobs because of automation.[35] This study by PWC has clearly identified the progress of the automation of jobs. It stretches over three overlapping times until 2030:

- *Algorithm wave* focused on automating simple computing tasks and analyzing structured data in areas such as finance, information, and communication—this is already well advanced;
- *Augmentation wave* focused on the automation of repetitive tasks such as form filling, communication and information exchange via dynamic technology support and statistical analysis of unstructured data in semi-controlled environments such as aerial drones and robots in warehouses. This is also underway, but is expected to mature in the 2020s;
- *Autonomy wave* focused on automating physical work and manual dexterity, and solving problems in dynamic situations requiring reactive actions, such as in manufacturing and transportation (e.g., driverless vehicles). Fully mature at the scale of the economy in the 2030s.

Statistical studies estimate that in more than 10 years the percentage of jobs lost to AI in the United States, Germany, the United Kingdom, and Japan will be 38%, 35%, 30%, and 21%, respectively in each country.[36] The SD redoubt the loss of employment.

The fear of the potential consequences of AI is not just about jobs, it extends to other areas. On the ethical level, there is a fear of divulging personal information manipulated by automated decision-making processes. The Montreal Privacy Statement has proposed, as part of its work, that the IA should offer privacy guarantees and allow individuals who use it to access their personal data as well.[37]

The analysis leads us to discuss the concept of security of information. Fundamentally, the security of information does not only concern virtual data. Also, palpable elements are concerned such as the physical security of the man and the natural environment because almost all human activities are insured by AI.

AI can potentially threaten the well-being and safety of humans by deliberately programming machines (robots) to harm. Autonomous machines are currently used for wars. In 2009, the United States used planes and autonomous cars in the Middle East. We may wonder if these machines were equipped with the decision-making capacity like humans, what would be the future of humanity.

The objective of researchers in the field of AI is to make machines with human intelligence. However, they forgot to rule on their legal responsibility. Concretely, the machine can be considered as a legal agent. In other words, can it be responsible for its actions before the law? Technically speaking, it can be because the law is a conventional tool for regulating social interactions and can, therefore, accommodate various legislative constructions, including the legal liability of autonomous artificial agents.[6]

Although a necessary and sufficient characteristic must exist in an agent to be legally responsible, the conscience, yet until further notice, machines are a set of electronic materials deprived of any form of "reason" and "sense."

Until now, intelligent machines are dependent on man and subject to his whims. They are only a sophisticated tool devoid of initiative impossible to empower them for acts dictated by man.

AI gives humanity a great satisfaction in terms of well-being. It has several advantages. Nevertheless, like any technological trend, it raises concerns about the possible negative effects. For this purpose, the SD carries out an environmental and societal monitoring about the activities using AI.

14.5 RECOMMENDATIONS

Technological advances have not stopped dazzling us to the point of wanting to make intelligent machines. Through the feats of AI, researchers want to create robots that would make our lives easier. The AI has covered all areas. It is not only for computers, but also for medicine, education, and even psychology.

The economy did not stay away from the invasion of AI's appetites. Since the advent of this technology, economic growth has been changing, particularly in developed countries. The scientific community does not universally appreciate the impacts of AI on the economy. Sustainable development researchers are reticent about the effects of AI. The questions raised by the SD about the effects of AI on humans and nature focus mainly on the consequences of job loss on the population, without neglecting the ethical and legal sides of the automation of the economy.

The author attempted in this chapter to explore the contribution of AI in achieving SD goals. AI has a proven record of accomplishment in various fields, particularly in economics. Several exploratory studies support this idea,[9,27,28,36,37] and believe that AI improves productivity and increases economic growth.

The author focused his research on the positive and negative impacts of AI on economic activity. The reticence of sustainable development is legitimate, but they must admit that AI can be at the service of SD.

Really, AI can positively respond to the environmental and social concerns of humanity. It allows access to education and health care, to fight against poverty and to eradicate hunger, and to fight against global warming. The major assets of AI are manipulation and processing of massive data. In medicine, for example, it offers citizens the means to become stakeholders in the health system. It allows diagnosing faster and more accurately and optimizing the search for new treatments.

Similarly, poverty and climate change will be detected through mapping. The automation of agricultural irrigation practices will revolutionize agriculture by strengthening food security. The skills of AI also provide for the rationalization of consumption of energies and natural resources as well as waste management.

Thus, the author believes that a relationship of complementarity exists between the two concepts. AI offers technologically sophisticated tools for achieving the goals of sustainable development. And the later keeps a watch on AI research to motivate them and direct them toward the consideration of possible adverse effects on humans and nature. AI inventions work together to achieve the goals expected by the SD because both aim at the well-being of humanity.

14.6 SUMMARY

The AI has covered all areas. It is not reserved exclusively for computing. It is used in medicine, teaching, and even in psychology. It has even upset the principles of the economy known since the 19th century. The IA in several sectors significantly influences economic growth: transport, logistics, supply chain, retail trade, information, communication, and finance. Opinions about the impact of AI on the economy differ. There are those, who believe in the AI's ability to drive economic growth. Others fear that AI affects the environment and the social well-being of the population.

However, only the countries holding this high technology will benefit. They will see their GDP increase by 2030 by about 14%. The labor market will recover because jobs will not be replaced but used more efficiently. In a SD approach, some reluctant people reproach to AI to ignore the negative impacts of AI-supported economic growth on people and nature. The SD denounces the impacts of AI on employment. They estimate that 37% of workers will lose their jobs. The SD also includes negative effects in the ethical and legal fields.

This study focused on the role that AI can play in achieving the goals of SD. AI, so apprehended by sustainable development, is making a massive contribution to reducing poverty and hunger, combating global warming, improving the health system, and many other areas. The AI is at the service of the SD despite the pessimistic opinions. The relationship between the two concepts gives the SD a right to look at all advances that the AI presents.

KEYWORDS

- **artificial intelligence**
- **externality negative**
- **externality positive**
- **sustainable development**

REFERENCES

1. Agnieszka, N. B. Introduction to Artificial Intelligence. http://zsi.tech.us.edu.pl/~nowak/bien/BIEN_introduction.pdf (accessed Jun 30, 2018).
2. Autor, D. H. Outsourcing at Will: The Contribution of Unjust Dismissal Doctrine to the Growth of Employment Outsourcing. *J. Labor Econ.* **2003,** *21* (1), 1–8.
3. Avenas, P. *Intelligence Artificielle (Artificial intelligence)*; p 32. https://www.lajauneet-larouge.com/sites/default/files/uploads/pdf-2017/733-page-032.pdf (accessed Jun 30, 2018).
4. Bellman, R., Ed.; *An Introduction to Artificial Intelligence*; Boyed Fraser Publishing Company: USA, 1978; p 157.
5. Bertrand, A. Émergence de l'intelligence artificielle. Quels défis et opportunités pour les PME bruxelloises et wallonnes (Emergence of artificial intelligence. What challenges and opportunities for Brussels and Walloon SMEs?). https://www.ucm.be/content/download/.../file/ucm-etude-intelligence-artificielle.pdf (accessed Jun 30, 2018).
6. Brozek, B.; Jakubiec, M. On the Legal Responsibility of Autonomous Machines. *Artif. Intell. Law* **2017,** *25* (3), 293–304.
7. Charniak, E.; McDermott, D. (Eds.). *Introduction to Artificial Intelligence*; Addison-Wesley Longman Publishing Co.: New York, USA, 1985; p 708.
8. Chris, S. *The History of Artificial Intelligence; Course CSEP 590A*; University of Washington: USA, 2006; p 27; https://courses.cs.washington.edu/courses/csep590/.../history-ai.pdf (accessed Jun 30, 2018).
9. Cüneyt, D. The Impacts of Robotics, Artificial Intelligence on Business and Economics. *Proc. Soc. Behav. Sci.* **2015,** *195*, 564–573.
10. Deloitte. *Artificial Intelligence Innovation Report*, 2016, p 40. https://www2.deloitte.com/content/dam/Deloitte/at/Documents/human-capital/artificial-intelligence-innovation-report.pdf (accessed Jun 30, 2018).
11. Frey, C. B.; Osborne, M. A. The Future of Employment: How Susceptible are Jobs to Computerization? 2013; p 72. https://www.oxfordmartin.ox.ac.uk/downloads/academic/The_Future_of_Employment.pdf (accessed Jun 30, 2018).
12. Guy-Warwick, E. *Artificial Intelligence: Where We Came From, Where We Are Now, and Where We Are Going.* Project Submitted in Partial Fulfillment of the Requirements for the Degree of Master of Science, University of Victoria: Canada, 2013; p 53.
13. Hindi, R.; Janin, L. Anticiper les impacts économiques et sociaux de l'intelligence Artificielle (Anticipate the Economic and Social Impacts of Artificial Intelligence); Report of the Working Group CNNum: Paris, France, 2017; p 48. http://www.strategie.gouv.fr/sites/strategie.gouv.fr/files/atoms/files/rapport-intelligence-artificielle-ok.pdf (accessed June 30, 2018).
14. Hindi, R.. 2015. https://medium.com/snips-ai/how-artificial-intelligence-will-make-technology-disappear-503cd88e1e6a (accessed June 30, 2018).
15. Jones, C. I.; Aghion, P.; Jones, B. F. Artificial Intelligence and Economic Growth. *Working Paper*, No. 17-027; Stanford Institute for Economic Policy Research (SIEPR). siepr.stanford.edu, 2017, p 56.

16. Joost, N. K.; Egbert, J. W. B.; Walter, A. K.; Van Der Putten, P. *Artificial Intelligence: Definition, Trends, Techniques, and Cases*. https://www.eolss.net/ebooklib/sc_cart.aspx?File=E6-44 (accessed Jun 30, 2018).
17. Khelfaoui, M. Corporate Commitment in Corporate Social Responsibility: Case of Algerian SMEs. PhD Thesis for Department of Economic Science, University of Algiers 3, Algeria, 2014; p 330.
18. La CCMM. *L'intelligence artificielle un pilier de développement économique pour la métropole (Artificial Intelligence a Pillar of Economic Development for the Metropolis)*; p 10. http://www.ccmm.ca/~/media/files/news/2018/intelligence_artificielle_pilier_developpement_economique_pour_la_metropole.pdf (accessed June 30, 2018).
19. Martin, J. Y. Environnement et Développement- Quelques réflexions sur le Développement durable (Environment and Development: Some Reflections on Sustainable Development)? *Doctr. Pract. Eval.* **2002,** 55–71.
20. Marwala, T. *Impact of Artificial Intelligence on Economic Theory*; University of Johannesburg: South Africa, 2015; pp 1–4. https://www.researchgate.net/publication/281486806_Impact_of_Artificial_Intelligence_on_Economic_Theory (accessed June 30, 2018).
21. Meghyn, B. Notes du cours "Introduction `à l'Intelligence Artificielle (*Introduction to Artificial Intelligence*)"; University, Aix-Marseille: France, 2009; p 89. https://www.lri.fr/~meghyn/papers/cours_IA.pdf (accessed Jun 30, 2018).
22. Meghyn, B.; Libert, B. Rise of AI Makes Emotional Intelligence More Important; https://hbr.org/2017/02/the-rise-of-ai-makes-emotional-intelligence-more-important (accessed Jun 30, 2018).
23. Miailhe, N.; Hodes, C. The Third Age of Artificial Intelligence. *J. Field Actions* **2017,** *17*, 6–11; e-article.
24. Mijwel, M. M. *History of Artificial Intelligence*; University of Bagdad: Iraq, 2015; pp 1–5. https://www.researchgate.net/publication/322234922_History_of_Artificial_Intelligence (accessed Jun 30, 2018).
25. Mijwel, M. M. Artificial Intelligence and Robot Technology; January 2018; online; https://www.linkedin.com/pulse/artificial-intelligence-robot-technology-maad-m-mijwel (accessed Jun 30, 2018).
26. Nilsson, J. Real-Time Control System Delays. Ph.D. Thesis for Department of Automatic Control, Lund Institute of Technology, Sweden, 1998, p 213.
27. Nilsson, N. J. Artificial Intelligence, Employment and Income. *AI Mag.* **1984,** *Summer* 5–14.
28. Parkes, D. C.; Wellman, M. P. Economic Reasoning and Artificial Intelligence. *Science* **2016,** *349* (6245), 267–272.
29. Pool, D.; Mackworth, A. K.; Goebel, R. Computational Intelligence: A Logical Approach. *Computacion y Sistemas* **1998**, *2* (2–3), 146–149.
30. Purdy, M. Artificial Intelligence Heralds Dramatic Potential for Growth for Both the Economy and for Humans; online. https://www.accenture.com/us-en/insight-artificial-intelligence-future-growth (accessed Jun 30, 2018).
31. Purdy, M.; Daugherty, P. Why Artificial Intelligence Is the Future of Growth. *Accenture* **2016,** 27. https://www.accenture.com/t00010101T000000Z__w__/fr-fr/_acnmedia/PDF-33/Accenture-Why-AI-is-the-Future-of-Growth.PDFla=fr-FR#zoom=50 (accessed Jun 30, 2018).

32. PWC. The Macroeconomic Impact of Artificial Intelligence. https://www.pwc.co.uk/economic-services/assets/macroeconomic-impact-of-ai-technical-report-feb-18.pdf (accessed Jun 30, 2018).
33. PWC. Sizing the Prize What Is the Real Value of AI for Your Business and How can you Capitalize? p 32. https://www.pwc.com/gx/en/issues/analytics/assets/pwc-ai-analysis-sizing-the-prize-report.pdf (accessed Jun 30, 2018).
34. PWC. Artificial Intelligence and Machine Learning; online. https://www.pwc.com/us/en/services/tax/tax-innovation/artificial-intelligence-and-machine-learning.html (accessed Jun 30, 2018).
35. PWC. Will Robots Really Steal our Jobs? An International Analysis of the Potential Long-term Impact of Automation; pp 47. https://www.pwc.co.uk/economic-services/assets/international-impact-of-automation-feb-2018.pdf (accessed Jun 30, 2018).
36. PWC. Highlights and Key Messages for Business and Public Policy; p 52. https://www.pwc.co.uk/economic-services/ukeo/pwc-uk-economic-outlook-full-report-march-2017-v2.pdf (accessed Jun 30, 2018).
37. Quebec Access to Information Commission. Pour un development responsable de l'Intelligence Artificielle qui respecte le droit à la vie privée et responsabilise tous les acteurs impliqués (For Responsible Development of Artificial Intelligence That Respects the Right to Privacy and Empowers All Involved); Montreal: Canada. https://www.declarationmontreal-iaresponsable.com (accessed Jun 30, 2018).
38. Rich, E.; Knight, K. *Artificial Intelligence*. 2nd ed.; McGraw-Hill Higher Education: New York, USA, 1990; pp 30–36.
39. Winston, P. H. *Artificial Intelligence*. 3rd ed.; Addison-Wesley Longman Publishing Co: New York, USA, 1992; pp 45–55.

CHAPTER 15

Mobile Technology Versus Era of Eco-Design

MOUNIA KHELFAOUI and SORAYA SEDKAOUI

ABSTRACT

Mobile technology (MT) has become the most widely used service worldwide, due to innovative applications it offers, from simple communications to the management of production processes in factories. MT also contributes to economic growth and social well-being. However, like any industry, it has been limited by ecological requirements due to principles of sustainable development. These constraints do not only concern the negative externalities of this industry on the man and nature, but also the whole process of manufacture of the products (and service) of the MT. The eco-design (ED) aims to integrate the environmental criteria of the design of a product (or a service) to reduce negative impacts at all phases of its life cycle. Therefore, this chapter has tried to answer the question: Will ED suffice to have an eco-responsible MT (ERMT)? In other words, whether innovations in MT contribute to achieving the goals of sustainable development (SD)?

15.1 INTRODUCTION

Internet access is increasingly being through smartphones and tablets, and computers and laptops are becoming almost obsolete. They are solicited for their practicality and their performance. Henceforth, information and knowledge are shared in very wide and expeditious manners. The benefits of mobile technology (MT) are numerous and infinite, they have facilitated our daily lives and allow us to perform painful tasks. In fact, "*Powerful mobile devices and sophisticated digital applications enable users to build businesses, access financial and health care records, communicate with public officials, and complete online transactions.*"[7]

The use of MT has changed the traditional relationships that people have with their professional activities. It equips them with the ability to exercise their professional activities anywhere and anytime.[5] MTs have transformed the structure of society by improving the level of education of individuals (MT tools are generally connected to the internet (via WIFI, 3G, 4G, etc.). They have also contributed to minimize social inequalities and to improve the status of women, especially in developing countries. Moreover, these participated massively in the economic growth of developed countries.

By taking advantage of the benefits of MT, people gradually lose their sociability. They also neglect the negative externalities arising from the use of MT. Indeed, the production process overexploits natural resources, in this case, the rare minerals (gold, silver, and copper); and generates considerable waste (knowing that these wastes can be recycled; a mobile phone consists of 500–1000 electronic components). For these reasons, the initiators of sustainable development (SD) and the proponents of technology have come together to sound the alarm on the possible harmful effects of modern MT on the human environment. Therefore, the advent of SD coincides with the publication of an article by two General Motors engineers, entitled "*Viable Industrial Strategies.*"[10]

In 1997, Hofman named "industrial environmentalism (industrial ecology)."[23] Subsequently, these notions have integrated the progression of the conceptualization of DD and have created the movement of "ED."

The company is no longer an isolated entity; it evolves in a stakeholder environment. The concerns (environmental, social, and economic) of these parties are an issue for its sustainability. It must, therefore, take into consideration from the conception of the product until its marketing.

Product design being the most important phase in the production process includes idea to realization. It is the decisive step, where the environmental aspects of products, raw material consumption, and environmental impact are determined.

Then eco-design (ED) or design for sustainability must include the pillars of the SD namely: the social, environmental, and economic elements of production.[20]

Apart from the much-praised benefits of MT, its products generate externalities that affect the environment and human health. It is imperative that they should be environmentally friendly products to assume the moral and physical responsibility of the outputs of their activities. Therefore, they must be fabricated with renewable resources, locally, without pollution, without too much waste and whose end of life does not require energy, and that they are biodegradable. In other words, these products must be eco-responsible (ER). Therefore, will MT sustainability initiatives be able to make MT products ER?

This chapter focuses on the concerns of SD community: Firstly, on the ability of the ED to make the MT ER; Secondly, in a general context if innovations in MT can contribute to the concretization of SD objectives.

The second section deals with the magnificent rise of the MT. The chapter also discusses the magnificent rise of the MT; the history of the evolution of the MT, by approaching its benefits and disadvantages for humanity. The conceptual framework of the ED has been analyzed. The relationship between the two ED and MT has also been presented. Authors also determined if there is a causal effect between the ED and the eventuality that TM can be ER.

15.2 EXTRAORDINARY MOMENTUM OF THE MOBILE TECHNOLOGY

The technical progress in MT offers the possibility to a multitude of applications in all domains of our lives. It has reshaped the behavior of people in society, facilitated access to information and knowledge, and has trivialized the tasks once very painful. "*Smartphonesand hand-held devices (HHD) combine advanced computing capability, such as internet communication, information retrieval, video, e-commerce, and other features, that make the device as one of the necessities for many persons. According to GSMA Intelligence, the number of mobile devices is 7.22 billion, while the US Census Bureau says this figure is still between 7.19 and 7.2 billion*" (https://www.independent.co.uk/life-style/gadgets-and-tech/news/there-are-officially-more-mobile-devices-than-people-in-the-world-9780518.html).

The evolution of the use of MT has no limit; it is a "phenomenon of society" as was the electricity at the end of the 19th century. People express an addiction to the range of tools offered by MT. Only one person can own up to 3–4 MT devices. Effectively, *The number of mobile subscribers around the globe has risen dramatically, from 2.3 billion in 2008 to 3.5 billion in 2014, and to 3.9 billion at the end of 2017.*[6]

The adventure of mobile technology dates back to the middle of the 19th century with the advent of wireless electronic transmission. Then, the term "cellular" came into being at the end of the 1940s without the evolution of technologies of its constituents (microprocessors). This was possible due to a succession of developments. It should be noted that the concept did not emerge in university laboratories but was rather inspired by very specific skills of the companies.[1] In general terms, the evolution of the wireless transmission, which is the basis of MT, has staggered the following steps:

1. *The step of wireless telegraphy*: It corresponds to the innovation by Marconi;
2. *The pre-cellular phase*: It comprises two stages: The emergence of wireless communications AM (amplitude modulation); and the transition to FM communications (frequency modulation) and MTS (mobile tele system); WiMAX was originally developed to deliver fixed wireless service with wireless mobility added in 2005. CDPD, CDMA2000 EV-DO, and MBWA are no longer being actively developed.
3. *The cellular phase* (https://en.wikipedia.org/wiki/Mobile_broadband; also called phase of "G" which counts four generations): The 1G from 1983 to 1992, that of analog cellular; The 2G from 1992 to 2001, that of the digital cellular; The 3G from 2001 to 2008, called cellular multimedia; the 4G from 2008 to the present day called broadband cellular; and the 5G from 2018.

The wide use of MT has advantages and disadvantages. However, the opportunities it opens up for people outweigh the negative consequences. Indeed, the MT participates massively in the economic growth.

The ICTs, by enabling productivity gains, promote economic growth and can explain the differences in performance observed between countries according to their degree of involvement in the "digital economy."[15]

Exactly, literature review on the link between mobile telephony and development has shown that an increase of 10 mobile penetration points would bring an additional 0.6–0.8 points of annual GDP growth,[17] due to wide distribution platform that MT offers to economic development. In a recent study, researchers have observed that for every 1% increase in penetration of mobile telephony, GDP may increase by 0.01%.[18]

In addition to the overall impact on GDP, a correlation has been established between foreign direct investment (FDI) and mobile penetration in host countries. For example, a 1% increase in the mobile penetration rate can result in 0.5–0.6% increase in the FDI/GDP ratio.[17] As well as the tax, revenues from foreign investment in mobile technology will make it possible to make expenses for the population.

It should be noted that the impact of MT on growth is lower in developing countries due to lack of ICT infrastructure. In addition, MT revenues in developing countries represent a significant percentage of GDP and an effective source for tax collection.[14]

However, "*the direct benefits of mobile phones in developing countries may include GDP growth, job generation, productivity gains, and increased taxable revenue. Employment is generated by both the mobile phone industry itself as well as the private sector, which provides the devices and airtime required to operate them.*"[18]

Thus, MT tends to digitize the structure of trade and the means of payment. There are several studies that have predicted the existence of four stages of mobile commerce[8]:

1. *The stage of digital commerce* characterized by the generalization of online purchases, applications or services (digital audio content, videos...) from mobile networks;
2. *The stage of generalization* of electronic commerce and all types of business (*business to consumer*) commerce via the mobile internet;
3. *The stage of the robot commerce* corresponds to the emergence and development of local commerce, including the payment of a person to an automaton;
4. *The stage of consumer-to-consumer payments*: money is transferred between individuals generally in the form of electronic wallets.

Certainly, this digital revolution has changed our lives it is enough for a single click, we can learn, distract, and even be served (remote control of domestic robots). Nevertheless, this technology has undesirable effects on family and social cohesion, on health and on the environment.

Paradoxically, when the person is connected to a smartphone or tablet, one is disconnected from reality for more than 3–4 h per day. Human contact is lost thus affecting moral and physical health.

The risks of health involved in using MT are numerous. Without being exhaustive, we can mention dependence, stress, musculoskeletal disorders, road risk, radiation by magnetic waves, and so on.[21] For example, a survey for Korean students showed that 73% respondents feel uncomfortable and are irritated when they do not have access to their smartphones. Similarly, a study of more than 3000 young people between the age of 14 and 16 found a significant association between smoking, alcohol use, and the intensity of smartphone use for both sexes.[13]

In the environmental component, the effects of MT are counter-balanced and oscillate between optimists and realists. For negativists, the use of large amounts of MT equipment generates pollution by consuming electricity. In addition, the manufacture of semiconductors still requires many solvents and chemicals that can contaminate the environment, although the industry has already made significant efforts to minimize negative effects. For example, hydrofluorocarbons (HFCs) and perfluorocarbons (PFCs) are potent greenhouse gases that contribute to the depletion of the ozone layer.[12]

At the same time, throughout its life cycle (from the extraction of first, through its manufacture, transport, use, and its end of life), the main environmental impacts of smartphones are depletion of resources, damage to biodiversity due to toxic discharges into the environment, and the emission of greenhouse gases.[2]

The manufacture of a smartphone (from ore extraction to final assembly) is responsible for about three-quarters of negative impacts, which are largely attributable to the screen and complex electronic components (microprocessors, etc.). The distribution and use of smartphone have less impact. The end-of-life stage has different impacts depending on whether the smartphone is recycled or not (https://link.springer.com/article/10.1007/s11367-014-0720-7).

Finally, MT is not immune to environmental concerns and must take steps to address the negative externalities generated by its manufacture and use.

15.3 THE CONCEPTUAL FRAMEWORK OF ECO-DESIGN

The various concerns on the environment have given birth to an ecological awareness since the 1970s, by denouncing industrial activity. Research has therefore focused on a new discipline that addresses the negative impacts of product design, to put in place an action plan promoting eco-technology. The latter has highlighted both the relevance of the concept ED and its maturity.[25]

The ED is the continuation of three approaches that have been operational in the last three decades. In the 1970s, the end-of-pipe approach was used to treat liquid, solid and gaseous tributaries. This same approach developed during the 1980s in a more preventive context is called "middle-of-pipe," to minimize waste and energy consumption. In the 1990s, this approach based on prevention was extended to the entire life cycle of products through the development of ED products.[19]

However, the traditional approaches to design differ from the current thinking of the life cycle in that it focuses on the intention of designers to include additional steps, such as upstream, downstream steps, and transport between each of them.[22] This reflection is deeper to integrate the negative externalities of production throughout the design process from the extraction of materials to recycling. Like all concepts, several reflections have tried to define it. The different definitions are listed according to five categories:

1. *Academic definitions* by academicians such as: Lindhal and Ekermann in 2013, Dewulf in 2013, Plouffe et al. in 2011, Borchardt et al. in 2011, Zhao et al. in 2013, and Pigosso et al. in 2010, Sherwin and Evans in 2000, Brezet and van Hemel in 1997, and Fiksel in 1996;
    - *Working definitions* correspond to those alleged by Sustainable scientists such as HOBE in 2011 and eco-design platform in 1996;
2. *Policy definitions* were formulated by European Commission in 2012, Eco-design Directive in 2005, PoleEco-conception in 2004, and OVAM in 2003;
3. *Standard definitions* are supported by ISO-14006 in 2011, EFA in 2003, and ISO-14062 in 2002;
4. *Encyclopedic definitions*: Bhamra and Lofthouse in 2007, Lexique in 2006, Fuad-Luke in 2002, and Wikipedia.

The concept of ED has been approached from several angles. However, all definitions agree that the ED aims to reduce environmental impacts, reflect the life cycle of the product and focus on product design. It should be noted that ED is a concept of SD and must respect the three components of sustainability: social, economic, and environmental.

Integrating the environmental impacts of a product into the production process is a new understanding of the product. It is about achieving technical innovations that can be made to meet the requirements of sustainability. On the economic level, it allows to create value, competitive advantage, and guarantee. Paradoxically, ED presents itself as a new constraint for the company.

Some research studies present their constraints as risks. They identify three risks related to ED: (1) environmental risk reflects the awareness of individuals and businesses of the damaging consequences for the environment of economic activity; (2) the risk related to the offer is that of the company for modifying its production conditions; and (3) market risk is linked to uncertainty about the market's response to seemingly identical but differently, designed products.[26]

The companies contribute to satisfying society's needs for goods and services. On the other hand, their activity has harmful effects on humans and the natural environment. Thus, the society revolted and denounced the actions of the companies. The requirements of society encourage companies to take responsibility for their negative externalities. Faced with these claims, they are obliged to review their production methods. ED is one of the steps companies take to align themselves with SD.

Businesses face unforeseen challenges, including the product's compliance with SD requirements and its ability to meet consumer needs. The company is facing a challenge of competitiveness on one hand and pressures of stakeholders on the other. As a result, it must develop strategies to manage eventual riskiness.

The purpose of ED is to commercialize products and services that respect the environment. These are products like those whose production process does not take into account environmental requirements. The problem is to convince the consumer to buy the eco-designed product, especially since its price is relatively high.

The company must anticipate countering the risks of the market. It is typically the mission of "Marketing" to develop a strategy to promote a new way of life to fit into the habits of consumers. For this, the marketing team relies on the promotion of a way of life ecological, more equitable and more responsible by targeting specific consumers: the ecologists, and the conscious ones. Then extend its action to a larger mass, laggards, and even polluters.[13]

The company is forced to convince the consumer with arguments like efficiency (consumes less energy), sustainability (designed to last longer), health, and local manufacturing (contributes to local development). However, the mission will only be accomplished if the two essential elements of the mix-marketing are relearned: the design and a competitive price.

ED remains an opportunity for companies beyond the burdensome considerations of environmental constraints because it will have financial advantages (cost reduction and profit) and reputation on the market.

Companies that are active in the field of TM are, more than ever, encouraged to adopt this mode of production to take advantage of the benefits of eco-accountability.

15.4 THE ECO-DESIGN: AN ASSET FOR THE ECO-RESPONSIBILITY OF MT OR SIMPLE ILLUSION?

The negative externalities leave an imprint on the environment due to the use of energy and raw materials, waste discharges (solid and liquid), and the emission of greenhouse gases. Also, the impacts of different productions are not confined to one place, but they spread beyond the border environment of companies. They vary according to the good or the service considered according to the raw materials, the processes of transformation and manufacturing processes, the design, technologies, and techniques used in manufacturing, the type of product, the packaging, the distribution method and the disposal method.[27]

MT products are particularly problematic objects for the environment in terms of waste flows. Indeed, they have many dangerous substances that can pollute the air when they are burned in incinerators; and leach through the soil and pass into the drinking water when they are buried in dumps.[4] "*Many of these toxic substances—including antimony, arsenic, beryllium, cadmium, copper, lead, nickel, and zinc—belong to a class of chemicals known as persistent toxins, which linger in the environment for long periods without breaking down. Some of them—including the metals lead and cadmium—also tend to accumulate in the tissues of plants and animals, building up in the food chain to dangerous levels even when released in very small quantities.*"[22]

As detailed here, the ED aims to integrate the negative externalities of production along its lifecycle. The studies have shown that 80% of the future impacts of the products are defined by decisions made during the design of the latter.[11]

To do this, several tools for analyzing and reducing the environmental impact are available to the company, including life cycle analysis (LCA), which is most recognized approach.[9,16]

When we refer to ED, we think directly of the product life cycle because this is the first step to achieve for this operation. ED differs from traditional design in that the first is initiated by the LCA approach. This approach makes it possible to quantify the impacts of a "product" (whether it is a good, a service or even a process), from the extraction of the raw materials that make it up to its elimination at the end of life, passing through the phases of distribution and use, or "from cradle to grave."[3,9]

The ED benefits from an international reference system, where these principles have been elaborated: ISO-14026 entitled "Environmental Management—Integration of Environmental Aspects into Product Design and Development." It concerns the methodology that ED is applied and based on several stages: planning, preliminary design, detailed design, testing (prototypes), production and market launch, and finally product review.

The life cycle approach aims to:

1. Obtain a global and exhaustive vision of the environmental impacts by reflecting by thinking in two ways: (1) *Multi-steps*: raw materials, manufacturing, transport/distribution, packaging, use, end of life (non-exhaustive list, to be adapted to the context); and (2) *Multipacks*: water consumption, energy consumption, pollution, waste production, and so on;
2. Prioritize environmental impacts according to their degree of importance and, if necessary, make "informed" arbitrations;
3. Understand that reducing these environmental impacts is a systematic process.

Once the environmental assessment of the product has been carried out, thanks to LCA, the company starts integrating ED into the production process as and when required. It calls on an ED strategy called the "methodological trojan horse" mechanism (mTh). It is an underlying strategy allowing for the introduction of a collaborative work style providing competence to the company and making it stand-alone on future eco-design programs.[27]

The state of the art has allowed us to identify three fundamental axes of the mTh[24]: (1) an eco-design tool adapted to the context of the company, a co-design approach for this tool in order to best translate the different constraints; (2) needs of users; and (3) organizational learning process based on the progressive mobilization of action levers in order to facilitate the integration of the environment in design.

To do this, the mTh is performed according to four principles: (1) the environmental assessment is done with the LCA; (2) after the LCA, the designers co-design the tool; (3) the stage of experimentation and validation of the tool; and (4) phase of harmonization of collective practices. Then, the mechanism ends with recommendations for eventual modifications.[24,27]

15.4.1 CASE STUDY

To assimilate an LCA, authors of this chapter have discussed a case study conducted by the company CODDE for ADEME in 2008 (Note: Veritas CODDE office is a center of expertise specializing in LCA and ED. For more than 20 years, it has been supporting manufacturers and distributors from all lifestyles in the implementation of sustainable design processes). This study concerns a mobile phone type "Barphone" 2nd generation, with an LCD screen of 15 cm^2. It is a phone without photo, video, and GPS functionality (only calls, SMS incoming and outgoing, directory, alarm clock, and calculator).

It is equipped with a standard external electric charger with a total mass of 338 g (133 g for the phone, 130 g for the packaging, and 120 g for the charger.) The manufacture and packaging of the phone are done in Asia (upstream transport is considered). As for distribution, it is operated in Europe by boat and then by truck. The phone is used for a period of 2 years, without consumables or maintenance parts, 5.5 h in and out by month. The charge of the telephone is 45 min in off-mode. No end-of-life treatment is envisaged (the telephone will end up at the bottom of a drawer).

The environmental impact assessment is carried out with software called EIME (version 0.3: EIME is a LCA software. It calculates the potential environmental impacts of a product or service from inventory data. Currently, it is developed and maintained by CODDE). The environmental indicators sought are the depletion of natural resources, primary energy consumption, the additional greenhouse effect, destruction of the ozone layer, acidification of the air, and eutrophication of water and production of hazardous waste.[9]

The result of this expertise revealed that the manufacturing phase is the most impacting and the elements responsible for the majority of the negative effects; and in order of importance are the LCD screen, the electronic set (except battery), the battery in lithium-ion, and the charger. The transport phase contributes very little to impacts on environmental indicators, and that of use affects the environment from 1% to 19% impacts.

It can be concluded that the most important step in the manufacturing phase, followed by the use phase; the transport phase is almost negligible. Thus, a scenario is planned to reduce the impacts due to the manufacture and use of this mobile phone.

Finally, the study is focused on a phone without great functionality and revealed significant impacts on the environment. The manufacturing of a smartphone or a tablet is very sophisticated that requires the miniaturization of the components. MT tools constitute a very great source of environmental problems. For this reason, they are challenged by the principles of ED more than other products and services to assume their ecological responsibility and become ER.

15.5 RECOMMENDATIONS

MT has metamorphosed our lives. The objects around us have become electronic gadgets competing with human intelligence. On the other hand, MT can help create and foster new opportunities for economic development. "Smart" economies can learn quickly through knowledge that is widely disseminated by MT. The latter has proved a fertile field for innovation in all fields, particularly those of the economy and the social. The need of the human in MT continues to increase, thus allowing increases its importance and its development.

The addiction of humanity to MT products has led to increased demand and consequently to negative impacts on nature and humans. Thus, natural resources are overexploited, uncontrolled waste, emanations of toxic gases and greenhouse effect. Also, several studies have shown the risks of TM on human health. They are psychic, physical, and even physiological.

Companies active in the field of MT are incited to integrate the prerogatives of sustainable development into their production strategies. They are expected to reduce the effects of negative externalities of their activities.

New disciplines less greedy in raw materials and energy are to be considered. The ED is a very good alternative to the "traditional" production methods that will participate in launching a dedicated SD era.

The objective of this chapter is to address the concerns of sustainable development in terms of internalizing the negative effects of MT manufacturing and use. In other words, is the ED able to make the MT responsible for its harmful actions toward the environment and man or ER?

This chapter has focused on the advantages and disadvantages of the intrusion of MT into our lives. Really, MT brings to humanity undeniable benefits. It contributes to economic growth, to improve productivity and the employment rate. It also allows us to inform ourselves and to distract ourselves.

On the other hand, the production and use of MT have undesirable effects on the natural environment, it destroys the family and social cohesion, and it even affects the health of a user. MT tools cause damage to the environment from the extraction of raw materials to its decline, "that is, its life cycle."

To minimize environmental stress, several approaches are recommended in the context of ED. The best of them is LCA, which aims to integrate SD in the process of production of the product from birth to the "grave." As a result, ED is the best alternative for integrating SD into the manufacture and use of TM products. It is likely to become an ER technology.

15.6 SUMMARY

The MT allows us to communicate without making physical efforts to move, to be quickly rescued in case of danger, to launch alarm calls in case of disasters and so on. It is also the most effective way of getting informed and cultivated. MT also contributes to economic growth. In fact, every 1% increase in the penetration of mobile telephony, GDP could increase by 0.01%. As well as the FDI/GDP ratio increases from 0.5% to 0.6% for a mobile penetration percentage of 1%. The benefits of MT on the economies of the developed countries, in particular, are felt in jobs, incomes, tax collection, and trade.

Certainly, the man has subjected the tools of the MT for his well-being. But, this enslavement is not without effects on its health, its natural environment, and its economy. The studies are much more alarming about the impacts of TM on human health and list certain diseases such as stress, dependence, musculoskeletal disorders, road risk, and radiation by magnetic waves.

The main environmental impacts of smartphones are the depletion of resources, damage to biodiversity due to toxic discharges into the environment and the emission of greenhouse gases. For all these reasons, the MT is forced to align itself with SD and to use one of the means offered by science in the field of sustainability, namely ED.

This study has tried to demonstrate the interest of ED for MT and its potential to give responsibility for the negative externalities it generates (manufacture and use). A case study of the environmental impacts of CODDE for ADEME, with the LCA approach, has revealed that the production phase is the most impacting, that of use is less impacting. While the effects of distribution are negligible. Thus, the use of ED for MT is imminent and is likely to make them ER. However, economically, ED products are not widely accepted by consumers, even though they are environmentally friendly. The commercial success of eco-designed products must take into consideration the product's aesthetics, profitability, and competitiveness. In other words, the ED must solve the paradox of the reconciliation between the economy and the environment.

KEYWORDS

- **eco-design**
- **eco-responsible mobile technology**
- **mobile technology**
- **sustainable development**

REFERENCES

1. Abaji Kharat, S.; Panage, B. M. History and Development of Mobile Technology with Evaluation of Mobile Library Services: Proposed Mobile Library Card (MLC). *Int. J. Lib. Inf. Sci.* **2015,** *4* (1), 9–23.
2. Agence de l'Environnement et de la Maitrise de l'Energie (ADEME). Les impacts du smartphone- Un téléphone pas si « smart » pour l'environnement (The Impacts of the Smartphone: A not so « Smart » Phone for the Environment); https://www.

ademe.fr/sites/default/files/assets/documents/guide-pratique-impacts-smartphone.pdf (accessed Jul 19, 2018).
3. Agence de l'Environnement et de la Maitrise de l'Energie (ADEME). L'approche en cycle de vie (The Life Cycle Approach). https://www.ademe.fr/sites/default/files/assets/documents/bao_achats_responsables_approche_cycle_vie.pdf (accessed Jul 21, 2018).
4. Bereketli, L.; Genevois, M. E.; Ulukan, H. Z. Green Product Design for Mobile Phones. *Int. J. Soc. Behav. Educ. Econ. Bus. Indus. Eng.* **2009,** *3* (10), 1–5.
5. Besseyre des Hors, C. H.; Isaak. H. L'impact des TIC mobiles sur les activités des professionnels en entreprise (THE impact of Mobile ICT on the Activities of Business Professionals). *Revue Francaise de Gestion* **2006,** *2006* (168–169), 243–263.
6. Brookings Institution. Mobile technology; 2016; p 2. https://www.brookings.edu/wp-content/uploads/2016/07/Chapter-One.pdf (accessed Jul 14, 2018).
7. Brookings Institution. Mobile Technology. https://www.brookings.edu/wp-content/uploads/2016/07/Chapter-One.pdf (accessed Jul 14, 2018).
8. Chaix, L. Le paiement mobile: modèles économiques et régulation financière (Mobile Payment: Business Models and Financial Regulation). *Revue D'economie Financiere*, **2013,** *2013* (112), 277–298.
9. Coulomb, J.; Feron, N.; Gonima, N.; Perrizato, C. Eco-conception: Analyse du cycle de vie d'un téléphone mobile (Analysis of Life Cycle of Mobile Telephones); https://moodle.insa-lyon.fr/.../ACV_telephone_portable_coulomb_feron_gonima_peri (accessed Jul 22, 2018).
10. Diemer, A. La technologie au cœur du développement durable: Mythe ou réalité (Technology at the Heart of Sustainable Development: Myth or Reality)? *Innovations*, **2012,** *2012* (37), 73–94.
11. Domingo, L. Methodologie d'éco-conception orientée utilization (Methodology of eco-design oriented use). Ph.D. Thesis, University of Grenoble, St. Martin D'Heres, p 178.
12. Faucheux, S.; Petit, O.; Hue, C. NTIC et Environnement: Enjeux, Risques et Opportunités (ICT and Environment: Issues, Risks and Opportunities). *CAHIERS DU C3ED*, **2001,** *1* (3), 2–30.
13. Grant, J. The Green Marketing Manifesto. https://www.wiley.com/en-ca/The+Green+Marketing+Manifesto-p-9780470687314 (accessed Jul 21, 2018).
14. Gruber, H.; Koutroumpis, P. Mobile Telecommunications and the Impact on Economic Development. https://www.isto.bwl.unimuenchen.de/download/forschung/ictcm/gruber_koutroumpis.pdf (accessed Jul 18, 2018).
15. Guédou, B.; Lelarge, C.; Blanchet. D ; Gallon, S. ; Lamotte, H. Impacts macro et microéconomiques des Technologies de l'Information et de la Communication Etat des connaissances en 2006, (Macro and Microeconomic Impacts of Information technologies and communication state of knowledge in 2006). http://www.crest.fr/ckfinder/userfiles/files/Pageperso/clelarge/docs/Impacts_macro_et_microeconomiques_des_Technologies_de_l_Information_et_de_la_Communication.Etat_des_connaissances_en_2006.pdf (accessed Jul 18, 2018).
16. Guibert, A. L'impact des outils d'éco-conception sur le degré d'innovation des enterprises (The impact of eco-design tools on the innovation capacity of business), RIODD; Saint-Étienne: France, 2016. hal-01350002 (accessed Jul 18, 2018).

17. Huet, J. M.; Viennois, I.; Labarth, P.; Kebede, T. A. La téléphonie mobile facteur de développement? (Mobile Telephony Development Factor?). *L'Expan. Manag. Rev.* **2012,** *137*, 118–127.
18. Hyde-Clarke, N. The Impact of Mobile Technology on Economic Growth Amongst 'Survivalists' in the Informal Sector in the Johannesburg CBD, South Africa. *Int. J. Bus. Soc. Sci.* **2013**, *4* (16), 149–156.
19. Mathieux, F.; Brissaud, D.; Zwolinski, P. Product Eco-design and Materials: Current Status and Future Prospects. https://www.researchgate.net/.../1752387_Product_ecodesign_and_materials_current_sta (accessed Jul 19, 2018).
20. Miakotko, L. The Impact of Smartphones and Mobile Devices on Human Health and Life. https://www.nyu.edu/classes/keefer/waoe/miakotkol.pdf (accessed Jul 17, 2018).
21. Passeron, J. Dépendance de la téléphonie mobile et risques associés. Revue de Littérature (Dependence of Mobile Telephony and Associated Risks: Literature Review), *References En Sante Au Travail*, INRS, **2016,** *145*, 131–139.
22. Prendville, S. Envisioning Eco-design: Definitions, Cases studies, and practices. https://www.researchgate.net/publication/301779042_Envisioning_Ecodesign_Definitions_Case_Studies_and_Best_Practices (accessed Jul 19, 2018).
23. Puel, G.; Fautrero, V. Ecoconception dans le secteur des TIC et relocalisation des activités : étude de cas dans des TPE françaises (Eco-design for IT Industry and Relocation of Activities: Case Study in French VSB). *Développement durable et territoires*, **2013,** *4* (3), 15–18.
24. Reyes-Carrillo, T. L'Eco-conception dans les PME: Les mécanismes du Cheval de Troie Méthodologique et du choix de trajectoires comme vecteurs d'intégration de l'environnement en conception (Eco-Design in SMES: Mechanisms of the methodological Trojan Horse and the Choice of Trajectories as Vectors Integration of the Environment in Design). Ph.D. Thesis, University of South Toulon-Var, France, 2007; p 237.
25. Roux, P.; Patingre, J. F.; Giroux, F.; Naud, O. Ecotechnologies et Ecoconception: concept et mise en œuvre (Eco-technologies and Eco-design: Concept and implementation). *Ingenieries* **2005,** *42*, 55–70.
26. Vernier, M. F. L'Ecoconception: une prise de risque pour l'entreprise? (Eco-design: A risk-taking for the company?). https://www.researchgate.net/.../272133579_L'Eco-conception_une_prise_de_risque_po (accessed Jul 20, 2018).
27. Villard, A. Développement de méthodologies d'Eco-conception pour le secteur microélectronique (Development of Eco-design Methodologies for the Microelectronics Sector). Ph.D. Thesis, University of Grenoble, St. Martin D'Heres, 2012; p 227.

CHAPTER 16

Reforming Public Finance Management in Africa Using Integrated Financial Management System

EMMANUEL EILU

ABSTRACT

Many countries in Africa still have weak public finance management (PFM) systems, and this has led to high corruption levels, poor service delivery, and high poverty levels. Reforms in PFM processes are critical if greater transparency and accountability in the management of public finances is needed. Many governments in Africa are still exploring better ways to promote accountability, transparency, efficiency, effectiveness, security of data management, and comprehensive financial reporting. Most common among these have been efforts to introduce the use of information and communication technology in integrated financial management systems (IFMS) for key aspects of PFM, such as in budget execution and accounting operations across all institutions of government. This chapter discusses the role of IFMS in improving PFM in Africa, its setback, cases and suggests ways of implementing a successful IFMS project in Africa. The lessons outlined in the chapter may be important in informing governments in Africa on how to successfully improve PFM using IFMS.

16.1 INTRODUCTION

In many transition and post-conflict countries in Africa, massive gains have been registered not only in democratic political governance but also in economic development. Politically, it is common knowledge today that many African countries are embracing democratic values in governance. For two decades now, competitive elections have become the standard in transition and post-conflict countries in Africa.[3] For example, nations such as Liberia, Mauritius, Sierra Leone, and South Africa have experienced massive leaps toward peace, justice, and economic development as the free and fair elections in these countries are conducted regularly.[3] There is evidence that many countries across the African continent have embraced democracy and are enjoying rapid economic growth and development.[7] For example, the average economic growth in the Sub-Saharan region (SSR) was about 6% in 2014, and 7 of the world's fastest 10 economies in 2011–2014 are from the SSR.[1] However, the pressures on governments to deliver basic services and improve the living conditions of their people are still enormous. Despite strides made in political and economic development, there are still colossal challenges in the public finance management (PFM) in many African countries, hence hampering effective services delivery and promoting corruption.

Generally, poor management of public finance in many African countries is seen as one of the major contributors to poor accountability and low levels of services delivery. Public financing in these countries is generally hampered by deep-seated deficiencies in PFM. Many countries in Africa do still have weak government financial control systems and this has maintained very high corruption levels, poor service delivery and high poverty levels in countries such as Burundi, Benin, Mali, and so on.[7] Many PFM systems used in number of African governments are not providing timely and accurate financial information for statutory reporting and decision making, especially in critical areas of budget planning and management, procurement, and asset management.[30] Currently, there is a wider call for reforms in PFM processes in different governments in Africa. Reforms in PFM systems and processes are critical if greater transparency and accountability in the management of public finances is needed. Transparent and efficient management of PFM means improved service delivery by governments, and citizens getting critical services (such as education, healthcare, and clean water).[29]

However, the nature and level of corruption risks occur at various stages of the PFM cycle, ranging from administrative to political corruption.[14] Although both administrative and political corruption require similar attention, yet many of the PFM reforms in African countries primarily address (and have a greater impact on) bureaucratic or administrative forms of corruption rather than political corruption.[14] Morgner[21] asserts that bureaucratic or administrative reforms in PFM such as improving the performance of the PFM system through improved budget execution and accounting tools are likely to have the greatest impact on (bureaucratic) corruption than political reforms. Though in the African context, political reforms are equally as significant in improving PFM as administrative reforms. This explains why in many African countries, donors support PFM reform projects that tend to primarily address and have a greater impact on bureaucratic forms of corruption rather than political corruption.[14]

Although a handful of African countries have successfully dealt with some of the bottlenecks in their PFM systems through some kind of reforms (such as in Burkina Faso, where momentum for budget reform was gained by involving many stakeholders like civil society in PFM reform committee). While Mozambique benefited from linking planning and budgeting, creating a good basis for a further push on reform. In Tunisia, progress was enabled by good coordination, consultation and a clear road map and sequencing of the reform. However, despite great efforts to reform PFM, many challenges still persist.[29] Many governments in Africa are still exploring better ways to promote accountability, transparency, efficiency, effectiveness, secure data, and a more robust means of fiscal accountability in PFM. International funding institutions like the World Bank, the International Monetary Fund (IMF), and USAID have supported and are still supporting and shaping PFM reforms through projects that provide assistance such as financial support, training, and technical expertise. Most common among these have been efforts to introduce the use of Information and Communication Technology (ICT) in strengthening PFM processes, such as in budget execution and accounting operations across all institutions of government. This is usually referred to as integrated financial management systems (IFMS).[2]

IFMS is the use of ICT by the government to perform PFM activities to foster efficiency, effectiveness, transparency, and accountability for public fund.[2,16] Good PFM reform in the form of IFMS supports not only good governance, political development, and transparency, but it is also critical for effectively delivering the services on which human and economic development relies.[12] For these reasons, many bilateral organizations and multilateral institutions consider PFM reform in form of IFMS to be a priority. For example, since 1984 the World Bank has financed different IFMS projects in different countries to the tune of $938 million.[8] From 2001 to 2010, external funding for PFM reforms in form of IFMS in Burkina Faso, Ghana, and Malawi was $35 million, $39 million, and $45 million, respectively.[12] Although much efforts have been made toward improving PFM using IFMS, yet not much is known on the extent IFMS addresses major challenges in PFM in Africa, and what best practices exist.

Therefore, this chapter reviews the state of PFM in the African continent, the reforms currently being implemented in form of IFMS, and the extent to which IFMS addresses significant challenges in PFM. This chapter also identifies the challenges of implementing IFMS in PFM in Africa. Finally, this chapter outlines number of key factors that are critical for the successful implementation of IFMS in Africa. The lessons outlined in this chapter may be important in informing governments in Africa on how to successfully improve PFM using IFMS to strengthen government financial control systems, improve service delivery, reducing corruption, and improving transparency and accountability in the national budget-making process.

Three major research questions are answered in this chapter as follows:

- What is the state of PFM in Africa?
- To what extent, do IFMS address major challenges in week PFM systems in Africa and what are the bottlenecks?
- What are the critical success factors for implementing IFMS in PFM systems in Africa?

16.2 METHODOLOGY

A narrative review was the main method used in this chapter. A narrative review provides a detailed summary of different primary studies from which conclusions are made.[4,18] One of the reasons this method was chosen was that it comprehends the diversities and pluralities around scholarly research topics and gives the writer an opportunity to write with self-knowledge, reflective practice, and acknowledgment of shared educational phenomena.[17] Secondly, a narrative review can address one or more questions and provides an overview of a broad spectrum of material in an easily readable format. It addresses the research questions by carefully selecting, chronicling, and ordering a park of evidence based on the research questions asked.[11,12] However, biases are the main weakness normally ascribed to narrative review due to subjectivity in study selection. In a narrative review, the study is not replicable.[12] However, given the nature of the entire chapter and its subsequent research questions, a narrative review analysis was the ideal method to address the research questions, because it gave comprehensive evidence within each of the research questions. The review involved two steps as follows:

- *Framing questions for a review*—This chapter comes up with clear research questions that guided the structure of the chapter. The following three particular questions guided the narrative review process:
  - *What is the state of PFM in Africa?*
  - *To what extent do IFMS address major challenges in week PFM systems in Africa and what are the bottle necks?*
  - *What are the critical success factors for implementing IFMS in PFM systems in Africa?*
- *Data collection*—An extensive and unbiased literature search was conducted on the role of biometric national identification technology in narrowing the identity gap in sub-Saharan Africa. The study selection criteria were directly guided by the research questions. Keywords were identified from the research questions. The keywords that were identified and searched for the first research question were: *"PFM in Africa," "major challenges of PFM systems in Africa."* Keywords searched from the second research question were: "*IFMS in Africa," "Opportunities of implementing IFMS,*" "*Challenges of implementing IFMS."* Keywords searched from the third research question included; "*Critical success factors for implementing IFMS.*" The review covered reports, journals, conference proceedings, books, and websites. Google Scholar was of much help for this review. The selection of papers was based on how each article comprehensively addressed each of the research questions and particularly the keywords.

16.3 INTEGRATED FINANCE MANAGEMENT SYSTEM

According to Berger,[2] for IFMS, ICT is used to systematically follow fiscal procedures and brings out a summary of fiscal transactions and activities. IFMS massively supports effective management and reporting of financial activities. In the government's context, IFMS refers to computerized PFM processes interlinked to different government departments, agencies, and other public-sector operations.[2]

IFMS does not only enable faster and efficient access to reliable financial data but it also strengthens government financial control systems, improves the provision of government services, raising the budget process to higher levels of transparency and accountability, and expediting government operations.[2] IFMS have number of components that work together to have an effective and efficient operation.

16.3.1 BASIC COMPONENTS OF THE GOVERNMENT IFMS

According to Berger,[2] basic IFMS consist of several distinct components or modules that use information already captured to execute different tasks. There are both simple and complex IFMS, the choice of particular IFMS depends on the task at hand. This chapter presents the basic components an IFMS should possess. Figure 16.1 presents components of a basic IFMS system.

FIGURE 16.1 Basic components of an IFMS.

Source: Adapted with permission from Ref. [2].

Given the scope and the nature of this chapter, it is not possible to explicitly explain all the components of an IFMS shown in the diagram above. However, one core component in the diagram is the *General Ledger*, or sometimes referred to as the *central database*. *The General Ledger* in many cases acts as the database of any IFMIS. All other modules post their transactions to the General Ledger. The standard Chart of Accounts is the custodian of all rules that govern all transactions in the IFMIS starting with the allocation of budget funds through to the commitments to pay for goods and services.

Many countries in Africa are in their initial stages of Implementing IFMS. However, few are in advanced stages of integration of IFMS in the entire PFM process. For example, South Africa has public accounting systems that are fully automated but not yet fully integrated across all tiers and departments of government. Uganda and Ghana have achieved a greater degree of integration, but do not yet have fully automated public financial accounting systems.[7] The IFMS is a large undertaking with potentially major benefits. Many of the countries that have implemented IFMS have greatly benefited.

16.3.2 BENEFITS OF IMPLEMENTING IFMS

- *Attract donor aid*: Many times, PFM reform in many cases in form of IFMS in Africa has been driven by the requirements of international donor organizations and foreign governments extending financial aid to countries on the continent. Many development partners required minimum standards of financial reporting to account for financial aid being provided to governments.[7] In 2005 Paris Declaration, as part of a global effort to make development aid more effective, partner countries especially those dependent on donor aid committed to strengthening their national PFM systems, while donors committed to using these systems to the maximum extent possible.[28] Number of donors want to see that the financial aid extended to a particular country is well spent through efficient and effective finance management systems and processes. With IFMS, governments are in a better position to borrow from international lenders on the open market, and at cheaper rates, to fund large infrastructure projects.[7] For example, countries such as Burkina Faso, Tanzania, and Zambia registered an increase in donor support following successful implementation of IFMS and overall PFM reforms.[20] Overall, African citizens stand to gain from such reforms, through greater public financial transparency and accountability, and the more effective use of public funds for service delivery.[7] An IFMS would improve the efficient use of donor aid as well as overall budget performance, and thus contribute to macroeconomic stability and growth. It would contribute to overall governance through the protection of public resources against the risk of expropriation and corruption.[24]

- *Integration of government systems and standardization*: IFMS integrates all government operations for using a common system across government institutions. In this system, many of the processes in government ministries and departments share the same integrated database. For example, accounting, cash management, commitment control, complementary payroll, budget preparation, aid and debt management, revenue administration, asset management, fleet management, and so on would all be managed under one centralized database and these different components would be communicating with each other. The integration of IFMS across the board ensures that all users adhere to common standards, rules and procedures, with the view of reducing risks of mismanagement of public resources.[5] Since governments are not money-making institutions, their IFMS are normally designed to ensure compliance with budget laws, other public finance rules and restrictions, and an entirely different set of accounting rules and reporting requirements.[2] For example, the IFMS used by the Kenyan government monitors and provides approvals for all activities and transactions from the beginning of the procurement process to the end. One of the most notable functions in this IFMS is the item "Master" responsible for standardizing the utilization of items within all Government Ministries, Department, and Agencies (MDAs), and counties. This effectively manages the price inflation of commonly used items in Government offices.[22]
- *Fighting Corruption*: Although Chene[5] claims that there are very few assessments of the impact of IFMS on corruption, yet there is evidence in the literature that IFMS has a potential to deter corruption to a certain extent. Robust IFMS provides mechanisms to track all financial transactions by the various departments that use the IFMS. The system can detect un-sanctioned transactions, under or overpayments, any form of fraud. The system can do this through automatic cross-checks for asset inventories to detect theft, automatic checks to detect failure to adhere to cash disbursement rules, automated checks for identification of ghost workers and so on.[5] The introduction of IFMS ensures that authorities can track every single transaction, which is extremely hard with manual

systems. For example, according to an expert, the manual systems cost the Kenyan government up to $1 billion in financial losses, and this is a value that could be brought down by 90% to $100 million with IFMS.[23] While in Nigeria, the government IFMS has not only helped reduce physical movement of money to make payments but also remove over 62,893 ghost workers in the government payroll systems, hence saving the government about $1 billion.[26] In Sierra Leone, the introduction of IFMS and subsequent reforms in PFM enabled government to uncover anomalies in personnel records of 2000 civil servants, leading to 16% of the subset employees being immediately suspended from the payroll because of the exercise.[5]

16.3.3 CHALLENGES OF IFMS

However, just like any other computerized system, IFMS also comes with challenges which, if not managed well, may lead to its failure. The implementation of IFMS has proven very challenging, especially for developing countries; and according to the World Bank's own account has not always been successful.[5] Out of 55 completed FMIS projects, 23 were in International Development Association (IDA) countries (12 completed in Africa). Almost 66% of the IDA funding ($747M out of $1133M) has been allocated to Africa. Despite these investments, Africa has the highest rate of failure in IFMS projects (4 out of 12 completed projects did not result in any operational PFM system), mainly due to initial attempts to implement ambitious IFMS solutions without adequate consideration of the limitations in capacity and infrastructure.[8]

In Uganda, for example, the current systems based on the Oracle Financials platform fairly work well; however, it is burdened with key design problems that will necessitate a system migration. A month after the Government of Uganda commissioned the IFMS, it was discovered that there were several deficiencies in the design of the CoA fields, leading to months of delays and considerable cost overruns. The deficiency required a rebuild of the system and this costed an additional $6 million. This would have meant going back to the World Bank to negotiate an increase in funding. These discrepancies arose because of inadequate planning, poor communication between implementers (donors and government), shortage of management capacity and resources, changes in system design without full agreement of all, poorly implemented trainings.[2,5,10]

While in Ghana, the Budget and Public Expenditure Management System (BPEMS) did not progress well, resulting in major implementation delays. The rollout of the system, originally planned for the end of 2001, was not achieved. The government was not satisfied with the BPEMS reporting system leading to major disputes between the government and the software team.[5,10,20] Nevertheless, despite challenges involved and many failed implementation attempts across the world and especially in Africa, there are a number of countries where IFMIS implementation is viewed as having been a relatively smooth and successful process such as Tanzania, South Africa, Ethiopia, and Burkina Faso.[5,16,20] Some of the strategies for successful implementation that other countries in Africa and developing countries, in general, can benefit from successful implementation of IFMS.

16.4 SUCCESSFUL IMPLEMENTATION OF IFMS

16.4.1 POLITICAL SUPPORT AND LEADERSHIP

Reforms in any sector of the economy will only be successfully executed when there is full support from the political leaders. According to Berger,[2] introducing and successfully implementing IFMS demands the government to commit to change toward the use of ICT. Coordination between the technical level and the political level is crucial and any form of disagreements should be settled immediately to avoid any escalations. Such reforms in PFM need to be "sold" to the politicians (including the opposition) especially when the political situation is fluid and changing.[20]

For example, Rwanda, Kosovo, Georgia, and West Bank and Gaza have achieved positive results within the framework of PFM reforms in form of IFMS because of the strong political commitment by the national governments.[21] While in the Slovak Republic, success in IFMS was because of real acceptance at the highest levels of the political system.[2] In 2005 in South Africa, political commitment was demonstrated by passing a cabinet resolution approving the implementation of IFMIS in government.[16] From 2004 to 2009, Malawi enjoyed strong political support for the implementation of IFMS, as was evidenced by the appointment of strong technicians as Vice President and Minister of Finance, and the sourcing of significant domestic funding for PFM reform, though that support has significantly reduced since 2009.[20]

16.4.2 AN INITIATIVE OF GOVERNMENT AND NOT DONORS

Although many African countries are in desperate need for donor aid to finance a bigger percentage of their national budgets, reforms in PFM should not be imposed as a donor condition. About a decade ago, $60 billion was earmarked as donor aid to flow to SSR to help these countries alleviate poverty and achieve the millennium development goals (MDGs).[24] However, this aid was normally tied to certain conditions that the recipient government must adhere to before aid fund is given. Many times, this condition has been tied to PFM reforms including implementing IFMS. It could be true that reform happens faster when tied to budget support, though it is more likely to be sustained if ownership is developed.[20] Many times, donors have often focused particularly on reforming the way budgets are drawn up yet managing the execution of those budgets has often fallen below standards. In other words, PFM reform of this nature may result in comprehensive planning for how donor money will be spent, but little by way of monitoring and ensuring that such funds are used effectively for the benefit of a country's people and economy.

According to Cosma,[7] PFM reform in Africa has always not been an initiative of the governments themselves but has mostly been driven by the need to comply with the requirements of international donors and development partners. However, many African governments have now realized the need for improving the management of their finances. If it is not the government's personal initiative, it is then a donor requirement. PFM reforms like IFMS is most likely to stop when the money stops. Therefore, PFM reform is a process that different countries will conduct in different ways, depending on their varying levels of maturity.[7,20]

16.4.3 CAPACITY BUILDING

For successful execution and sustenance of an IFMS, there is need to build considerable human resources throughout the entire government departments and ministries. According to Chene,[5] the low level of computer literacy in developing countries especially in Africa must first be adequately addressed before such projects can be truly adopted and implemented. Building capacity to successfully execute and sustain IFMS requires a cross-sectional restructuring. From technical and administrative training to review of salary structure, attractive enough to compete with private sector employment conditions and to incentivize candidates with required IT skills.

In capacity building, the different user groups must be identified, their levels of knowledge determined, recruiting needs established and training curricula explored. Training programs need to cut across the various audiences, from senior members of the bureaucracy down to mid- and entry-level civil servants.[2] For example, the National Treasury of Kenya has stepped efforts to enhance the adoption of its online PFM system by training more than 5000 users at IFMS Academy based at the Kenya School of Government. The 5000 trainees came from both national and county governments.[22] In Uganda, the Ministry of Finance and Economic Development recruited a team of 12 Ugandan experts to support the implementation of the IFMS and its capacity-building program. The 40 graduate staff specifically trained to support the system were also recruited, many of which have since been absorbed into the civil service. The internal training facility that was set up during the project was key to developing the skills.[27]

16.4.4 PLAN BIG AND IMPLEMENT SMALL

Starting with pilot schemes is ideal, as errors can easily be identified and rectified in this early stage. Although it seems best to design the whole IFMS, yet it is important to implement piece by piece. It is important to consider the breadth, speed, and depth of the IFMS project. African experience suggests an in-depth holistic designing of the whole IFMS project but recommends a gradual implementation based on priorities and human capacity, which is consistent with the real needs and the capacity to implement.[20] Besides that, there is a need by all government structures to begin using standard or common platforms or tools. According to Berger,[2] implementing an IFMS requires different government ministries, departments, and agencies to begin working with a common tool. There is a need for information consistency across all the administrative units in local and regional governments. The units have to adopt a common "language" in the form of a uniform Chart of Accounts (CoA). Therefore, it is paramount to have all PFM units relevant to the IFMS participate in the design and implementation. In Uganda, for example, government units with key responsibilities in budgeting (for budget preparation and releases, and sector planning), macro-fiscal units, accounting, internal audit, procurement, and external audit were all standardized and used a common platform.[27] While in Vietnam, more than 5 years were required to reach agreement on a common CoA.[2] Given the enormousness of an IFMS project, Hendriks[16] observed that a phased kind of implementation has a better chance of success because the projects can easily be monitored and any issues that may arise from the implementation can easily be addressed. The IFMS in South Africa, for example, was implemented in phases, from phase one which is designing the overall master plan, to phase two—which is developing capacity, and phase three was to implement the master plan. Pilot schemes are carried out both nationally and provincially.[16] While in Tanzania, the roll-out plan was incremental in nature, where the first implementation processes began at the Accountant General's Department and 10 pilot ministries. After a consolidating successes in this phase, the system was later rolled out to all 43 ministries and departments in the capital and then rolled out further to the entire central government.[22]

16.4.5 APPROPRIATE CHOICE OF TECHNOLOGY

There is a wide disparity between developed countries environment and the African environment. The African environment particularly provides a greater challenge that calls for a redesign of ICT to appropriately suit African structures. Morgner[21] stresses that the effectiveness of IFMS solely depends on the robustness, flexibility, and appropriateness of the technological solution chosen. The technology chosen should not only be administratively appropriate, but also its social, cultural, and political relevance is equally important. There is a need to appropriately align the IFMS technology with the existing environment especially the structures, policies, and laws of the country. Through all this, Berger[2] observed that the existing systems must be addressed as part of the assessment needs so that data can easily be transferred from the legacy systems into the future IFMS. In Uganda, for example, the concept of open technology architectures was adopted, which allowed flexible implementation of interfaces and integration with other systems for sharing data and enhancing efficiency. The IFMS can easily interface with systems at the Uganda Revenue Authority (URA) and the Bank of Uganda (BoU) and interface with the integrated personnel and payroll system (IPPS) at Ministry of Public Service.[27] As opposed to the simple appropriate design suggested by the World Bank and its consultants, the so-called "state-of-the-art" model opted by senior government officials in Ghana was a massive failure because of inappropriate project model which did not blend with the existing structures. By 2010, despite an investment outlay more than $20 million, the system was not operational in any of the eight pilot areas.[20] Unlike the IFMS in Ghana, the IT solution selected in Tanzania was a medium-sized management and the accounting package was significantly less complex.[21]

16.4.6 CHANGE MANAGEMENT

Change management ensures there is a flawless transition from the legacy system to the future system. To achieve this, it is necessary to attend to the wider impacts of introducing IFMS to government, and as well as considering the tangible impacts of change, ranging from personal impact of IFMS on those affected, to getting accustomed to working and behaving in new ways. Morgner[21] observed that many times IFMS reforms have been introduced as instruments for a radical and comprehensive restructuring of procedures to jumpstart and improve the PFM and this result into resistance from those affected. Those with selfish interest, or those who know that IFMS will constrict their means of corruption will resist change toward IFMS.[16] When IFMS was introduced to government ministries in Uganda, there were entrenched practices or power networks that were purportedly lined up for termination because of IFMS, coupled with requirements to learn new skills and the fear of layoffs among staff, created negative resentments' toward IFMS. However, an extensive change management exercise and a communications strategy were implemented to minimize the fears and negativity, and to improve ownership. These were accompanied by a set of incentives to motivate staff to use the IFMS.[27] While in Ghana, the IFMS project design was criticized for particularly paying little attention to change management and to the assessment of capacity constraints and training needs, and also for being technology driven.[20]

16.4.7 LEGISLATIVE AND LEGAL FRAMEWORK

Enacting laws that govern the IFMS is crucial for its success. Ratifying clear guidelines, rules, procedures, and roles that structure the whole process, from implementation to maintenance of IFMIS is very important.[16] The IFMIS implementation must be supported by a comprehensive legal structure by which the overall public finance system is based.[5] In Tanzania, for example, in 2001/2002, 3 years since the adoption of IFMS as a framework and system of managing public funds, a new and up-to-date legislation called the Public Finance Act (2001) came into force and this had the effect of synchronizing practice with the law.

16.5 SUMMARY

Despite huge steps made in political and economic development, there are still colossal challenges in the management of public finances in many African countries for using ICT. Many countries in Africa do still have weak government financial control systems and are not providing timely and accurate financial information for statutory reporting and decision making, especially in such critical areas as budget planning and management, procurement, and asset management. Currently, there is a wider call for reforms in PFM processes in different governments in Africa. Reforms in PFM systems and processes are critical if greater transparency and accountability in the management of public finances is needed. Many governments in Africa are still exploring better ways to promote accountability, transparency, efficiency, effectiveness, security of data management, and comprehensive financial reporting in PFM.

Most common among these have been efforts to introduce IFMS in key aspects of PFM such as in budget execution and accounting operations across all institutions of government. This was done as one of the most common PFM reform practices, aimed at the promotion of efficiency, effectiveness, accountability, transparency, security of data management, and comprehensive financial reporting in number of African countries such as Ghana, Uganda, Nigeria, Tanzania, Ethiopia, South Africa, and Burkina Faso.

IFMS do not only enable faster and efficient access to reliable financial data but it also strengthens government financial control systems, improves the provision of government services, raising the budget process to higher levels of transparency and accountability, and expediting government operations. In Africa, countries such as Tanzania, Ethiopia, South Africa, and Burkina Faso have successfully implemented IFMS and have greatly benefited from it. While countries like Uganda, Nigeria, and Ghana have registered partial failure and have not fully benefited from IFMS.

KEYWORDS

- **Africa**
- **integrated financial management systems**
- **public finance management reforms**
- **information and communication technology**
- **Sub-Saharan region**

REFERENCES

1. Amadou, S. Y. *Four Instruments to Strengthen Financial Integration in Sub-Saharan Africa*; Africa Growth Initiative: Brookings Institution, 2014; p 25.
2. Berger, L. *Integrated Financial Management Information Systems: A Practical Guide*; Louis Berger Group, Development Alternatives, Inc.: USAID, Washington DC, United States, E-book, 2008; p 118.
3. CAE (Colloquium on African Elections). Best Practices and Cross-Sector Collaboration: Final Report, 2009. http://www.ndi.org/files/African_Elections_Best_Practices_ENG.pdf (accessed Nov 1, 2018).
4. Campbell Collaboration. Systematic Reviews in Social Sciences. A presentation on February 7, 2006. https://campbellcollaboration.org (accessed Aug 31, 2018).
5. Chêne, M. The Implementation of Integrated Financial Information Management Systems (IFMIS). Online, U4 Helpdesk, Transparency International, 2009; p 11. https://www.u4.no/publications/the-implementation-of-integrated-financial-management-systems-ifmis/pdf (accessed Aug 31, 2018).
6. Cooper, H. M. The Integrative Research Review: A Systematic Approach. In *Applied Social Research Methods Series*; Sage: Beverly Hills, CA, 1984; Vol. 2, p 23. http://pareonline.net/pdf/v14n13.pdf (accessed Aug 31, 2018).
7. Cosma, J. The Rewards of Reform: Public Financial Management Reform in Africa. ED 0914, 0115 South Africa, 2014; p 16. https://www.ey.com/Publication/vwLUAssets/EY-The-reward-of-reform/%24FILE/EY-The-reward-of-reform.pdf (accessed Aug 31, 2018).
8. Dener, C.; Watkins, J. A.; Dorotinsky, W. L. *Financial Management Information Systems 25 Years of World Bank Experience on What Works and What Doesn't*. World Bank Study; The World Bank: Washington DC, 2011; p 28.
9. DFID (Department for International Development). *Systematic Reviews in international Development*, 2013. https://www.gov.uk/government/publications/systematic-reviews-in-internationaldevelopment/systematic-reviews-in-international-development (accessed Aug 31, 2018).

10. Diamond, J.; Khemani, P. *Introducing Financial Management Information Systems in Developing Countries.* IMF Working Chapter; IMF World Bank: Washington DC, 2005; p 29.
11. Dixon-Woods, M.; Agarwal, S.; Jones, D.; Young, B.; Sutton, A. J. Synthesizing Qualitative and Quantitative Evidence: A Review of Possible Methods. *J. Health Serv. Res. Policy* **2005,** *10* (1), 45–53.
12. Ferrari, R. Writing Narrative Style Literature Reviews. *Eur. Med. Writ. Assoc.* 2015, E-article, DOI: 10.1179/2047480615Z.000000000329.
13. Fjeldstad, O. *Anti-Corruption Reforms: Challenges, Effects and Limits of World Bank Support. Background Chapter to Public Sector Reform: What Works and Why?* Report No. 7; IEG Evaluation of World Bank support, World Bank: Washington DC, 2008; p 71. https://www.cmi.no/publications/3134-anti-corruption-reforms-challenges (accessed Nov 1, 2018).
14. Fjeldstad, O. *Taxation and Development: A Review of Donor Support to Strengthen Tax Systems in Developing Countries*. WIDER Working Chapter No. 2013/010; UNU- Wider (United Nations University World Institute for Development Economics Research): Helsinki, Finland. https://www.wider.unu.edu/publication/taxation-and-development. (accessed Nov 1, 2018).
15. Hemingway, P.; Brereton N. What is a Systematic Review? Hayward Group Website. April 2009. http://www.whatisseries.co.uk/whatis/pdfs/What_is_syst_rev.pdf. Published April 2009 (accessed Nov 1, 2018).
16. Hendriks, C. J. Integrated Financial Management Information Systems: Guidelines for Effective Implementation by the Public Sector of South Africa. *SA J. Inf. Manag.* 2012, *14* (1), 9. http://dx.doi.org/10.4102/sajim.v14i1.529 (accessed Nov 1, 2018).
17. Jones, K. Mission Drift in Qualitative Research, or Moving Toward a Systematic Review of Qualitative Studies, Moving Back to a More Systematic Narrative Review. *Q. Rep.* **2004,** *9* (1), 95–112.
18. Kirkevold, M. lntegrative Nursing Research An Important Strategy to Further the Development of Nursing Science and Practice. *J. Adv. Nurs.* **1997,** *25*, 977–984.
19. Kowalczyk, N.; Truluck, C. Literature Reviews and Systematic Reviews: What Is the Difference? *Radiol. Technol.* **2013,** *85* (2), 110–115.
20. Lawson, J. *Evaluation of Public Financial Management Reform Burkina Faso, Ghana and Malawi 2001–2010*. Final Synthesis Report, 2012; p 3. https://www.sida.se/globalassets/global/about-sida/sa-arbetar-vi/utvardering/evaluation-brief-evaluation-of-public-financial-management-reform.pdf (accessed Nov 1, 2018).
21. Morgner, M. Public Financial Management Reforms in Developing and Post Conflict Countries. U4-Expert Answer, 2013; p 13. https://www.transparency.org/files/content/corruptionqas/383_Public_financial_management_reforms_in_developing_and_postconflict.pdf (accessed Nov 1, 2018).
22. Mutegi, L. *Kenya Government Procurement Finally Moves*, 2014. https://allafrica.com/stories/201408140047.html. http://www.cio.co.ke/news/main-stories/kenya-government-procurement-finally-moves-online (accessed Nov 1, 2018).
23. Okutoyi, E. IT Web Africa: Kenya Turns to Tech to Fight Tender Corruption, 2014. http://www.itwebafrica.com/ict-and-governance/256-kenya/232482-kenya-turns-to-tech-to-fight-tender-corruption (accessed Nov 1, 2018).

24. Prakash, T.; Cabezon, E. *Public Financial Management and Fiscal Outcomes in Sub-Saharan African Heavily-Indebted Poor Countries*. IMF Working Chapter; World Bank: Washington DC, 2008; p 31.
25. Folscher, A. African Experience of Budget Reform. *OECD J. Budg. Spec. Issue Collab. Afr. Budg. Reform Init.* **2006,** *6*, 9–24.
26. Prime Time. Nigeria Lacks Institutions, Systems, to Prevent Corruption; Okonjo-Iweala, 2015. http://www.premiumtimesng.com/news/more-news/177045-nigeria-lacks-institutions-systems-to-prevent-corruption-okonjo-iweala.html (accessed Nov 1, 2018).
27. Semakula, L.; Muwanga, R. *Budget Strengthening Initiative: Uganda: Implementing an Integrated Financial Management System and the Automation of the Budget Process*; 111 Westminster, Bridge Road: London, SE1 7JD UK, 2012; p 65.
28. World Bank. *Using Country Public Financial Management Systems; A Practitioner's Guide*. 4th High Level Forum on Aid Effectiveness: Busan, Korea; World Bank, Washington DC, 2011; p 54.
29. World Bank Institute. *Public Financial Management Reforms in Africa*, 2012. http://wbi.worldbank.org/wbi/news/2013/05/13/public-financial-management-reforms-africa (accessed Nov 1, 2018).
30. Uganda–MFPED (Ministry of Finance Planning and Economic Development, Uganda). Integrated Financial Management System, 2012. http://www.finance.go.ug/mofped/integrated-financial-management-system (accessed Nov 1, 2018).

Index

A

Activity–tool–goal approach, 6–7
  conceived relationship between, 7
Adova Soft, 239
Africa
  mobile networks in agriculture, 50–54
    Chipata Women's Mobile SMS project, 51
    Dokoza, 52
    Mali and Burkina Faso, 52
    Maluleke project, 52
    South African Depression and Anxiety Group (SADAG), 52
    Uganda, 52
    Village Phone Initiative, 52
    Women of Uganda Network (WOUGNET), 52
    Xam Marsé SMS market information service, 52
  mobile technology
    role in combating illegal timber logging in, 91–99
  social media ecosystems
    BlueWorld, 66–67
    NaijaPal, 66–67
  social media in socioeconomic development, role, 61–71
    old *versus* new, 63
African social media ecosystems, 65–67
  challenges, 67–69
    currency and sociocultural poverty, 67–69
  Mxit, 65–66
  technological dependency, 69–71
Airtime
  webradios, platforms of, 220
Artificial intelligence (AI)
  economic growth, 252–253
  recommendations, 256–258
  review, 249–251
  role in sustainable development, 247–257, 253–256

B

Blogs
  social media platforms
    benefits in agricultural, 47
BlueWorld
  African, social media ecosystems, 66–67
Bottom of the pyramid (BoP), 4, 12–14
Broadcast Myself
  platforms of webradios, 220–221

C

Chipata Women's Mobile SMS project
  Africa
    mobile networks in agriculture, 51
Combating illegal timber logging in Africa
  mobile technology (MT), role, 91–99
    Cameroonian, 95
    discussion, 98–99
    forest watch, 98
    methodology, 94
    rainforest connection, 95–97
    results, 94–98
    TIMBY, 97
Community radios (CRs), 209–210, 213–215
Consumer relations management (CRM), 192
Country-specific brand crowd, 242

D

Digital social media, 177
  healthcare, importance in, 182–183
    advance applications, 182–183
  usage in health sectors, 178–180

Disruptive power of mobile technology (MT)
debates, 5–6
ecological transformation, 19–20
economic transformation, 19
Indian agro-supply chain, 25–37
potential disruptive power, 12–16
potential transformative power, 17
social transformation, 17–19
water resources management, 11–12
Dokoza
Africa
mobile networks in agriculture, 52

E

e-learning in adult education
WhatsApp, 119–132
background, 121–123
limitations and recommendations, 130–132
literature review, 123–126
materials and methods, 127
results and discussion, 127–130
uses, 124–126
Eco-design in mobile technology (MT)
asset for eco-responsibility, 271–274
conceptual framework, 268–271
era of, 263–275
Internet access, 263–264
recommendations, 274–275
Ecological transformation, 19–20
Economic transformation, 19
eMix, 183
Exclusiveness social media
luxury brands, 235–236

F

Facebook
social media platforms
benefits in agricultural, 47
Fear of getting copied, 242–243
Flickr
social media platforms
benefits in agricultural, 48

G

Goal, 9–11
MDGs
water resources sustainability, 11
ranked score, 10
Gross domestic product (GDP)
Sub-Saharan African (SSA), 75

H

Health systems
developing countries, promising approaches, 183–189
community or groups, 187–188
electronic data entry, 187
electronic education, 187
electronic mail, 184
individuals, 188–189
mobile devices apps and tasks, 185
social networking, 186
Web surfing, 186
developing nations, 181
mobile use, 189–194
"big data" in e-health, 193
connecting people, 191
consumer relations management (CRM), 192
convergence, 193
e-health, 192–193
emerging trends, 194
expanding information, 191–192
social media, 189–194, 194–197
"big data" in e-health, 193
assisting customer assessments, 195–196
awareness creation, 196
challenges and future directions, 199–200
connecting people, 191
consumer relations management (CRM), 192
convergence, 193
discussion, 197–199
e-health, 192–193
emerging trends, 194
expanding information, 191–192

importance, 194–197
major issues, 196–197
specialized skills recruitment, 195
Horizontal information exchanges (HIE), 12
potential disruptive power, 13
Human activities, 7–8

I

iMedical, 183
Indian agro-supply chain
transformative power of mobile technology, 25
applications and services, 31–32
challenges, 27–28
challenges faced by mobile app providers, 33–35
ecosystem, 29–30
future prospects, 36–37
information, 28–29
methodology, 27
opportunities, 35–36
policies, 30–31
requirements, 33
Information
potential in agricultural scenario, 42–43
sharing in agriculture context, 43–45
Information Communication Technology (ICT), 120
Institute of Extra Mural Studies (IEMS), 159, 160
Integrated financial management systems (IFMS), 279, 281, 283
basic components of, 284–285
challenges of, 287–288
implementation of
capacity building, 289–290
change management, 292
implement small, 290–291
initiative of government, 289
legislative and legal framework, 292–293
millennium development goals (MDGs), 289
plan big, 290–291
political support and leadership, 288
technology, appropriate choice of, 291–292
implementing, benefits of
attract donor aid, 285–286
fighting corruption, 286–287
government systems and standardization, integration, 286
Kenyan government, 287
iWander, 182

K

Kayunga district, 75

L

Lake Victoria Basin (LVB), 3
disruptive and transformative power of mobile phone in, 11–20
potential disruptive power, 12–16
water resource management in, 11–12
Tanzania, map, 12
Lesotho College of education (LCE), 159, 160
Lesotho Distance Teaching Center (LDTC), 159
modes of content delivery, 160
objectives, 160
Lesotho's distance learning
current distance learning, 167–169
integrating social media
distance learning, 169–171
gaps and opportunities, 157–172
limitations, 171–172
literature review, 162–167
material and methods, 167
results and discussion, 167–171
LinkedIn
social media platforms
benefits in agricultural, 48
Luxury brands
social media, 225–244
big apparel brands, 238
brand humanization, 234
brand socialization, 238–239
brand synchronization, 229–230
build brand equity, 234
challenges, 241–243

choice, 232–234
concrete loyalty, 234
connect long-term audience, 234
consumers, 240
correlation between, 227–230
customization, 234
gaps in, 231–232
mass marketing, 240
preferred social platforms, 227
public relations, 235
relationship capital formation, 235
social sync, 235
strong brand recognition, 228–229
virtual gap, 230

M

Mali and Burkina Faso
Africa
mobile networks in agriculture, 52
Maluleke project
Africa
mobile networks in agriculture, 52
Massive open online courses (MOOC), 141
Millennium development goals (MDGs), 9, 289
water resources sustainability goals, 11
Mobile and social media use
in health systems, 189–194
“big data” in e-health, 193
connecting people, 191
consumer relations management (CRM), 192
convergence, 193
e-health, 192–193
emerging trends, 194
expanding information, 191–192
Mobile app-based expert system in Uganda, 75–88
AI/expert system, 79
current status of tomato cultivation, 78–79
methodology, 80
objectives, 77–78
process model, 80–86
capture, image, 80–81
classification, 85
feature extraction, 84
implementation, 86
preprocessing, image, 81–83
segmentation, 83
testing classifiers, 85–86
results and discussions, 86–88
Mobile learning, 106
apps
role of teacher–developer relationship, 137–153
development of application, 145–147
Google research volume around, 156
operational use, 105–115
data reliability and validity, 112
descriptive statistics, 112–114
discussions, 115
hypotheses, 106–107
literature review, 108–111
research methodology, 111
results, 111–114
theoretical framework, 107–108
pedagogy transformation and applications, 142–145
recommendations, 152–153
teacher–developer
importance of collaboration, 147–152
Mobile networks in agriculture
Africa, 50–54
Chipata Women’s Mobile SMS project, 51
Dokoza, 52
Mali and Burkina Faso, 52
Maluleke project, 52
role in combating illegal timber logging in, 91–99
South African Depression and Anxiety Group (SADAG), 52
Village Phone Initiative, 52
Women of Uganda Network (WOUGNET), 52
Xam Marsé SMS market information service, 52
Uganda, 52
Mobile technology (MT), 3
combating illegal timber logging in
Africa, role, 91–99
Cameroonian, 95
discussion, 98–99

forest watch, 98
methodology, 94
rainforest connection, 95–97
results, 94–98
TIMBY, 97
disruptive power
debates, 5–6
ecological transformation, 19–20
economic transformation, 19
Indian agro-supply chain, 25–37
potential disruptive power, 12–16
potential transformative power, 17
social transformation, 17–19
water resources management, 11–12
eco-design
asset for eco-responsibility, 271–274
conceptual framework, 268–271
era of, 263–275
Internet access, 263–264
recommendations, 274–275
extraordinary momentum, 265–268
information
sharing in agriculture context, 43–46
learning, 139–142
networking and agriculture, 46
potential in agricultural scenario
future, 41–57
technology integration in learning
process
history of, 140
transformative power
debates, 5–6
ecological transformation, 19–20
economic transformation, 19
Indian agro-supply chain, 25–37
potential disruptive power, 12–16
potential transformative power, 17
social transformation, 17–19
water resources management, 11–12
usage, 3–4
Multiple phone functionality, 211–221
CRs, 213–215
discussions, 221–222
mobile phone
reinforcement of radio transmission,
215–216
streaming, 216–218
webradios, platforms of, 218–221
Airtime, 220
Broadcast Myself, 220–221
SAM broadcaster, 220
Mxit
African, social media ecosystems, 65–66

N

NaijaPal
African, social media ecosystems, 66–67
Non adoptability of popular selling
techniques, 241–242

O

Operational use
mobile learning, 105–115
data reliability and validity, 112
descriptive statistics, 112–114
discussions, 115
hypotheses, 106–107
literature review, 108–111
research methodology, 111
results, 111–114
theoretical framework, 107–108

P

Popular e-tailing brands, 227
Potential disruptive power
LVB, 12–16
HIE, 13
VIE, 13, 15
mobile devices, 16
trends and aggregation, 16
Potential transformative power, 17
Principle of on-demand
overview and alertness, 4–5
Process model
mobile app-based expert system in
Uganda, 80–86
capture, image, 80–81
classification, 85
feature extraction, 84
implementation, 86
preprocessing, image, 81–83
segmentation, 83
testing classifiers, 85–86

Public finance management (PFM)
systems, 279
IMF and USAID, 281
methodology, 282
data collection, 283
framing questions for, 283
Mozambique benefited, 281
public financing, 280

S

SAM Broadcaster
platforms of webradios, 220
Small-scale FFO
goal, 9–11
Smart and e-agriculture, 49
Social media
conceptual definition, platforms, 63–65
ecosystems in Africa
BlueWorld, 66–67
NaijaPal, 66–67
and exclusiveness
big apparel brands, 238
brand socialization, 238–239
brands, 240
challenges, 241–243
consumers, 240
mass marketing, 240
information, platforms
sharing in agriculture context, 43–46
luxury brands, 225–244, 236–238
big apparel brands, 238
brand humanization, 234
brand socialization, 238–239
brand synchronization, 229–230
build brand equity, 234
challenges, 241–243
choice, 232–234
concrete loyalty, 234
connect long-term audience, 234
consumers, 240
correlation between, 227–230
customization, 234
gaps in, 231–232
mass marketing, 240
preferred social platforms, 227
public relations, 235
relationship capital formation, 235
social sync, 235
strong brand recognition, 228–229
virtual gap, 230
platforms, 46–48
blogs, 47
Facebook, 47
Flickr, 48
LinkedIn, 48
online *versus* offline for luxury brands, 243–244
potential in agricultural scenario, 41–56
socioeconomic development role in Africa, 61–71
Twitter, 48
WhatsApp, 48
YouTube, 48
potential reasons, 239
Social networks
role in healthcare for developing countries, 177–200
Social transformation, 17–19
mobile tools, 18
South African Depression and Anxiety Group (SADAG), 52
Africa, mobile networks in agriculture, 52
Sub-Saharan region (SSR), 280

T

Transactional distance theory (TDT), 105
Transformative power of mobile technology
Indian agro-supply chain, 25
applications and services, 31–32
challenges, 27–28
challenges faced by mobile app providers, 33–35
ecosystem, 29–30
future prospects, 36–37
information, 28–29
methodology, 27
opportunities, 35–36

policies, 30–31
requirements, 33
TuDiabetes, 182–183
Twitter
social media platforms
benefits in agricultural, 48

U

Uganda
Africa
mobile networks in agriculture, 52
mobile app-based expert system in, 75–88
AI/expert system, 79
classification, process model, 85
feature extraction, process model, 84
implementation, process model, 86
methodology, 80
objectives, 77–78
process model, image, 80–86
segmentation, process model, 83
testing classifiers, process model, 85–86
tomato cultivation, current status, 78–79
mobile app-based expert system in
process model
preprocessing, image, 81–83

V

Vertical information exchanges (VIE), 12
potential disruptive power, 13
Village Phone Initiative
Africa
mobile networks in agriculture, 52
Virtual gap, 230
methodology, 231

W

Webradios
multiple phone functionality, platforms of, 218–221
Airtime, 220
Broadcast Myself, 220–221
SAM broadcaster, 220
WhatsApp
e-learning in adult education, 119–132
background, 121–123
limitations and recommendations, 130–132
literature review, 123–126
materials and methods, 127
results and discussion, 127–130
uses, 124–126
social media platforms
benefits in agricultural, 48
WhatsApp messenger, 123–124
Women of Uganda Network (WOUGNET), 52
Africa
mobile networks in agriculture, 52

X

Xam Marsé SMS market information service
Africa
mobile networks in agriculture, 52

Y

YouTube
social media platforms
benefits in agricultural, 48